The International Toxic Waste Trade

Christoph Hilz

VNR VAN NOSTRAND REINHOLD
_____ New York

Copyright © 1992 by Van Nostrand Reinhold

Library of Congress Catalog Card Number 91-28987
ISBN 0-442-00713-2

Manufactured in the United States of America

Published by Van Nostrand Reinhold
115 Fifth Avenue
New York, New York 10003

Chapman and Hall
2-6 Boundary Row
London, SE 1 8HN, England

Thomas Nelson Australia
102 Dodds Street
South Melbourne 3205
Victoria, Australia

Nelson Canada
1120 Birchmount Road
Scarborough, Ontario M1K 5G4, Canada

16 15 14 13 12 11 10 9 8 7 6 5 4 3 2 1

Library of Congress Cataloging-in-Publication Data

Hilz, Christoph, 1960–
    The international toxic waste trade / by
Christoph Hilz.
        p.   cm.
    Includes index.
    ISBN 0-442-00713-2
    1. Hazardous waste management industry.   2. Hazardous waste
management industry—Government policy.   3. Hazardous wastes—
Environmental aspects.   4. Hazardous wastes—Law and legislation.
I. Title.
HD9975.A2H55   1992
338.4'7363728—dc20   91-28987
                                                            CIP

To my parents

Contents

CHAPTER 8. Conclusions   172

CHAPTER 9. Recommendations   179

Foreword

Hazardous waste has become a problem of increasing importance for both industrialized and developing countries. As nations increase their industrial base, hazardous waste has been an expected by-product of economic activity. Ironically, in developed countries the generation of hazardous waste has been exacerbated by earlier attempts to reduce air and water pollution. The control of hazardous waste has come relatively late in the set of regulatory initiatives instituted in industrialized countries. As a result there has been a "media shift" resulting in a transfer of toxics from air and water to the hazardous waste stream.

Waste has been called the *visible* face of industrial inefficiency. Either disposing of those wastes in landfills or exporting them has constituted an attempt to render the problem *invisible*. However, the problem has now reached enormous proportions and new solutions are needed. These solutions require a re-examination of national, regional and international policies related to industrial production, hazardous waste management and the export of hazardous wastes.

Toxic wastes have well-documented human health and ecological consequences. Exporting those wastes can cause damage not only in the importing country, but can also be recycled to the exporting country in the form of agricultural and industrial products produced abroad, as well as through transboundary movement of pollution. Countries which accept hazardous waste may enjoy short-term economic benefits, but those countries may incur long-term serious ecological, health, and economic costs related to cleanup or restoration of their own environments and public health.

The export of hazardous waste violates the principles of sustainable development. Furthermore, exporting hazardous waste raises serious questions of fairness and distributional equity for the countries of import and transit. Finally, the ability to export hazardous waste provides a valve for exporting country to escape the pressures for toxic use reduction by externalizing the costs of production onto other countries. Thus, while some might view the issue of export of hazardous waste as tangential to problems related to the direct pollution caused by industrial and agricultural processes, addressing this problem is in fact central to the development of sound environmental policy.

In this book, Christoph Hilz traces the origins and nature of hazardous waste generation and export, and provides a long-overdue analysis of national and international policies affecting hazardous waste export and transportation. Specific recommendations are offered to alleviate this serious global problem, which represents the more visible tip of the iceberg of environmental pollution and degradation.

In analyzing the problems of continuing past waste practices, Christoph Hilz uses five criteria: environment and development, efficiency, risk, equity, and sustainability. In formulating recommendations for desirable future directions, he considers these factors as well as the practicality of various policies. Considered essential are redefining national sovereignty to allow international inspection, strengthening international protocols and information/technology transfer mechanisms, and creative use of liability systems. Source reduction and pollution prevention are argued to be the preferred approaches to minimizing waste. Significant infrastructure improvements in both legislative frameworks and management systems will have to accompany these changes. In this treatise, Christoph Hilz gives us much to ponder and provides us with a challenge to find a better way to sustainable and equitable development.

Nicholas A. Ashford
Associate Professor of Technology and Policy
Massachusetts Institute of Technology

Preface

April 22, 1990, marked the twentieth anniversary of the first Earth Day. At the end of these 20 years, the world community finds itself threatened by numerous environmental problems. Increasing amounts of hazardous wastes is one of them. Our capability to manage these wastes in an environmentally sound manner is severely constrained, and some countries face a state of "emergency." As a result, the export of hazardous wastes to other industrialized, and to developing and newly industrialized, countries has quickly increased. This book analyzes policies for controlling transboundary movements of hazardous wastes and shows that existing policies are insufficient for meeting the requirements of *sustainable* development. In order to achieve a sustainable path in our hazardous waste policy, concrete recommendations are put forward to strengthen the only international regulatory approach—the Basel Convention on the Control and Transboundary Movements of Hazardous Wastes and their Disposal.

Part 1 examines the underlying causes of hazardous waste exports, describes the institutional arrangements of the current practice, estimates the global scope and scale, and outlines the consequences of such a practice. The focus is to examine the intrinsic technical, economic, and sociopolitical factors of the waste export practice. Faced with growing production of hazardous wastes and limited disposal and processing capacity, costs for hazardous wastes management increased sharply over the last decade. Wide sensitivity and greater environmental awareness in the population, coupled with tighter environmental legislation, put additional constraints on domestic waste management. In the developing world, some nations' generally weak economies, exacerbated by large foreign debts, made these countries prone to accept waste imports in order to earn badly needed revenues.

Hazardous waste exports have caused serious detrimental consequences. Damage to the environment and ecology, threats to human health and life, additional economic costs for cleanup, and political and diplomatic difficulties have resulted from improper waste management abroad. For industrialized countries, waste exports directly implicate policies for waste minimization.

Part 2 provides an in-depth analysis of all available policy options to control hazardous waste exports and imports, by examining their economic effectiveness, equity balance, sustainability, and feasibility for implementation, in order to assess their merits and weaknesses. Policy options include the existing practice in industrialized countries, the developing world, and Eastern Europe; unilateral initiatives, as undertaken by the United States; international policies, such as bilateral and regional agreements, or a global ban; and various other initiatives pursued by the OECD, the EEC, and the United Nations. The purpose of this examination is to make concrete recommendations for a sound policy on transboundary waste movements, including waste management in LLDCs and NICs.

This book is important, as it provides a thorough analysis of the consequences of hazardous waste export policies. As such, it serves the fundamental dialogue in designing policies to control future developments in the hazardous waste management business, such as the export of hazardous waste management facilities. Already, the export of entire waste management facilities to developing countries is in the middle of the debate. Although a different activity, as it is argued that state-of-the-art technology would be exported and would ensure safe waste management, hazardous waste export to these facilities is anticipated and would thus raise similar economic, risk, sociocultural, and ethical questions.

This book is a response to many heated discussions and to the negotiations, on various national, regional, and international levels, on transboundary waste movements. So far, no comprehensive analysis on this subject has been provided, and the still existing differences over a common global policy show a need for this book. Its purpose is to help overcome these differences and to find a sustainable policy.

The International
Toxic Waste Trade

Part I

Establishing the Facts

1

The Need for a New Methodology

In the past several years, hazardous waste management policies have been adopted by almost all industrialized countries, in order to manage the hazardous by-products of industrial processes that have largely contributed to our affluent life. Today, many of these countries find themselves in an "environmental emergency", resulting from limited disposal and treatment facilities, strong public objections against new facilities, and increasing production of hazardous wastes. While in the past hazardous wastes were traded mainly among industrialized countries, in recent years they have been exported increasingly to developing countries.[1] These practices have not only caused an international scandal, but have brought into question the effectiveness of existing national policies for managing and controlling hazardous wastes. As a result of discussions and debate at the national, regional, and international levels, a search was initiated for an adequate policy. In addition, negotiations were sponsored by several international organizations, such as the Organization for Economic Cooperation and Development (OECD) and the United Nations Environment Programme (UNEP).[2] The need for a global policy on hazardous waste management is accentuated by the growing recognition that the economies and environments of industrial and developing nations are inextricably linked and rely on each other for natural resource inputs, as well as for markets.

Analyzing national, bilateral, regional, and international approaches for controlling transboundary movement of waste requires a methodology that goes beyond existing ones. The methodology developed in this chapter consists of a set

---

[1]International debt and economic problems in developing countries exacerbate the need for hard currencies. Importing hazardous waste provides an avenue to obtain currency.

[2]See discussion in Chapter 7.

of criteria that will be used for the policy evaluation. They include effects on *environment* and *development, efficiency, risk, equity,* and *sustainability.* Finally, there is the concern of *realpolitik* (i.e., whether the policies and solutions can be rendered operational)

ENVIRONMENT AND DEVELOPMENT

Development and the environment are interdependent in all nations. In the developing world, two considerations are of primary importance. First, economic underdevelopment, frequently linked to high foreign debts, is the single largest cause for environmental degradation, such as soil erosion, desertification, deforestation, lack of fuelwood, and clean water (Tolba 1987, 104). Second, many countries emphasize industrial development as the only opportunity to strengthen their economies and thereby improve living conditions. However, given their limited resources, environmental protection is often unaffordable and thus neglected. At the same time, increasing industrial production contributes additional pollution and hazardous wastes which stress the ecosystem. In 1972, The United Nations Conference on the Human Environment stated that, without development, there is no basis to allocate resources for environmental protection and improvement of living conditions, and without environmental protection, proper development is not feasible (Tolba 1987, 97).

In light of the interconnections between industrial and developing nations on one hand, and the environment and development on the other hand, it appears that a global policy on hazardous waste management is superior to isolated and uncoordinated policies that have resulted in numerous problems. Such a policy will have to satisfy a variety of criteria, in order to be feasible and acceptable to the world community. A global policy arguably must be efficient in allocating scarce resources. It must also address the ethical concerns that result from impoverished conditions prevalent in many parts of the world. Sustainability provides an additional criterion for policy design.

EFFICIENCY

The trade of hazardous waste should not be treated as another aspect of world trade in commodities, to be entirely unregulated or included in bilateral and international trade agreements, such as GATT[3]. However, besides many other complex differences, the uneven standards in waste management among the various countries create economic distortions. Exporting hazardous waste results in higher environ-

---

[3]General Agreement on Tariffs and Trade (GATT). See also discussion in Chapter 7.

mental quality in the country of export, while the costs for proper waste management are externalized to the importing country. As a result, from a microeconomic perspective of an exporting country, the practice of waste export is efficient. Large sums of money may be saved, by avoiding waste treatment costs, such as incineration, landfilling, or recycling, and liability costs resulting from possible future damages due to inadequate waste management, and insurance coverage for hazardous waste facilities. Distortions resulting from different regulatory standards, externalities, and the transfer of risk support the argument that hazardous wastes should not be handled as *economic goods*.[4]

The efficiency of exporting hazardous wastes derived on the basis of a cost-benefit analysis depends both on the time frames and the scope of the analysis.[5] In the short term, exporting hazardous wastes can be efficient because exporting countries save resources that would otherwise have to be spent for safe waste management and for ensuring public health. As far as importing nations are concerned, they can earn foreign currencies and other infrastructural benefits. Short-term benefits might well be of higher value than the anticipated risks and costs for waste disposal.

In the long term and on a global scope, hazardous waste exports could turn out to be very inefficient for both importing and exporting countries. Importing countries face many risks that they might not be able for control and that might cause irreparable ecological and public health damage. The future costs, to mitigate possible accidents and environmental damage and to ensure public safety, could outweigh the previous economic benefits many times over. Given the experience of many industrial countries where hazardous waste–related accidents and discovery of uncontrolled waste sites have occurred many years after the initial disposal, importing highly toxic wastes may pose a substantial economic burden. Likewise, industrial countries may reevaluate their economic analysis when they are confronted with unanticipated public health costs that result from diseases caused by imported agricultural products containing hazardous chemical residues. The case of pesticides, sprayed in developing countries and found in tropical fruits sold in industrial countries, illustrates the case (Weir and Schapiro 1981). The costs of possible damage may be considerably higher than the cost of proper disposal in the country of origin in the first place. Considering the complexity of the economic factors and the short- and long-term implications, the traditional interpretation of efficiency has severe limitations as a policy criterion.

---

[4]A meeting of GATT rejected a proposal by Nigeria to draft a protocol restricting the trade of hazardous wastes within GATT member states, pointing out that HWs (hazardous wastes) were not an issue of trade. See Chapter 7.

[5]The discount rate or time rate of preference is also crucial. See Ashford 1981.

RISK

Closely related to costs and benefits are the risks associated with hazardous waste management. Chemical substances or heavy metals as part of hazardous waste can alter the chemical, biological, or physical conditions of the ecosystem in a way that adversely affects life processes. Other possible sources of risk are the occupational safety and health risks associated with the production processes that generate hazardous waste, and waste transportation over long distances. In order to design a policy, the questions to be asked are: what sort of risks exist, how can they be quantified, what can be done to minimize, reduce, or otherwise manage them, and what level of risk is acceptable to society and the world community? Risk assessment depends primarily on factors, such as physical characteristics of the disposal site, level of toxicity of the wastes, the flora and fauna, and the extent of human exposure. But other factors, such as institutional assumptions, the legal and regulatory framework, and social and cultural values, are also part of risk evaluation (Whynne 1989, 134). People evaluating policy options for hazardous waste management will have to consider both the technical and the societal factors and find a balance between them. As a result, it might be necessary to extend the standard definition of risk, in order to satisfy all significant factors.

If exporting hazardous waste poses risks for public health, we need to examine to what extent we are willing to accept adverse effects on human health and life. A risk analysis on transfrontier movements of hazardous wastes could result, in different results depending on the country for which it is performed. Some developing countries may value the short-term benefits higher than the marginal increase of risks, as they perceive them. In any case, countries that import hazardous wastes must weigh the degree of risk against the relative social and economic benefits from the economic goods traded or profits earned.[6]

EQUITY

A general feature of transboundary pollution problems is that one country receives a variety of social and economic benefits, resulting from the production process, while the risks and dangers are shared by several countries, as water or air pollution crosses national borders. Hazardous waste exports are deliberate practices that can be controlled, regulated, or prohibited by national legislation, in contrast to transfrontier air or water pollution, which has been uncontrollable since nation states have no direct influence on pollution generated in other countries (Kelly 1985, 95). This raises questions about whether hazardous wastes movements are compatible with our moral and ethical values and our concept of responsibility. Hazardous

---

[6]For a discussion of trade-off analysis in the context of technology transfer, See Ashford 1981.

wastes that are finally disposed of abroad carry almost all the potential environmental and health impacts to importing countries, and, as indicated earlier, countries of transit might also be confronted with environmental risks if accidents occur. Most hazardous waste is generated in industrial countries. However, exporting waste results in potential risks primarily to people in importing countries, who do not share in the benefits of the waste generating production processes. The people who share the potential risks have little, if any, practical influence on the decision to import these wastes (WCED 1987, 35).

The export of hazardous wastes to countries with weaker environmental legislation and less stringent regulation constitutes a de facto double standard. In industrial countries, hazardous wastes are regulated to some extent.[7] Developing countries may not have adequate resources, sophisticated environmental legislation, or the administrative and institutional infrastructure to tackle the problems of hazardous waste management. Considering the discrepancy between environmental standards in industrial and developing nations, we need to question whether a double standard in environmental protection is acceptable or tolerable (U.S. Congress 1989, 341). The result is less protection for people in the countries of import and would clearly be a violation of the Stockholm Declaration,[8] which states in Principle 21:

> States have, in accordance with the Charter of the United Nations and the principles of international law...the responsibility to ensure that activities within their jurisdiction or control do not cause damage to the environment of other States or of areas beyond the limits of national jurisdiction (United Nations 1973; UNEP 1988, 7).

The distribution of wealth within and among different societies needs to be considered when evaluating policies for hazardous waste exports. We also need to appreciate the economic position that many developing countries have found themselves in at the beginning of the last decade. With deteriorating economies during the 1980s many developing countries are faced with major foreign debts that heavily burden their economies and socioeconomic development. In light of the desperate need for financial resources and the lucrative offers of hard currencies, in exchange for hazardous waste imports, an increasing number of countries were considering this trade. Industrial countries need to question whether it is morally justifiable to export hazardous wastes if another country agrees to import these wastes because of its poor economic position and great demand for financial

---

[7]For an analysis of hazardous waste management in selected industrialized countries, See OECD 1985.

[8]*Declaration of the United Nations Conference on the Human Environment.* June 16, 1972. The Stockholm Declaration was adopted by the United Nations Conference on the Human Environment, in Stockholm in 1972 (United Nations 1973).

resources. Questions of geographic equity are thus further complicated by questions of intergenerational equity (Brown 1981, 361). We also need to consider that the current generation has far greater potential to do irreparable harm to the environment than have any of its predecessors.

Equity is an important criterion for evaluating policies that might result in the moral and ethical problems and dilemmas described in the previous paragraph. The notion of equity refers to the principle of treating all people *equally* and in a *just* way (Rogets II 1980, 331). What do *equally* and *just* mean in a world with tremendous cultural, social, and economic differences? Equity can best be attained through a flexible approach that recognizes these differences, rather than through strict application of the laws of a given country. Principle 23 of the Stockholm Declaration seems to support this:

> ...it will be essential in all cases to consider the systems of values prevailing in each country, and the extent of the applicability of standards which are valid for the most advanced countries, but which may be inappropriate and of unwarranted social cost for the developing nations (UNEP 1988, 8).

The concept of equity includes ethical norms. One fundamental principle of ethics is *responsibility* (Jonas 1979), and a general interpretation is to preserve nature and life. Martin Honecker formulates in his essay, "Responsibility for the Environment as a Commitment for Humankind," (Honecker 1980) an ecological and environmental ethic in which responsibility implies three rules: (1) it is ethically reprehensible to transform the environment in a way that foreseeably damages present or future generations, (2) the environment can be used to meet the needs of humankind, as long as it does not cause future damage, and (3) the environment is to be preserved and protected (Honecker 1980, 13–15). These three criteria give us a guideline of factors to be considered in formulating a hazardous waste management policy and its sanctioned transboundary movements. They also imply a fairly new criterion—the concept of sustainability.

SUSTAINABILITY

Sustainability (Tolba 1987; WCED 1987; UNEP 1988; Brown 1981; Carr 1988; Stivers 1981) is a general term subject to a variety of definitions.[9] Basically, sustainable development refers to two broad concepts. The first implies using our resources in a manner that promotes "development that meets the needs of the present without compromising the ability of future generations to meet their own needs" (WCED 1987, 8). The second refers to the increasing environmental crisis

---

[9]A list of such definitions was compiled in a report by the World Bank (World Bank 1989).

and implies that any human activity should protect the global ecosystem—flora, fauna, and humankind (Jonas 1979, 22). These two concepts may appear as two separate issues, but they ought to be considered as being related. Environmentally sound management of wastes should become an integral part of resource management and resource safeguarding.

In recent years, hazardous waste production has steadily increased, reflecting both economic growth and increased use of resources. Current control strategies are designed to manage the wastes produced, rather than to avoid producing hazardous materials. As long as "inexpensive" disposal alternatives are available, there is little incentive to look for new directions, such as reducing waste at its source (Underwood 1988, 30). Policies that support these practices are not sustainable because the increasing environmental costs are not fully taken into account. The costs associated with hazardous waste management are often higher than the costs to prevent waste production. A sustainable policy would emphasize reducing waste to the minimum level possible and would effectively manage the unavoidable residual (Roelants du Vivier 1988, 80).

Sustainability also requires that we use our limited resources efficiently, giving emphasis to recycling and reuse. A report by the U.S. Environmental Protection Agency (EPA) stated more than a decade ago that treating hazardous waste was only the third choice after source reduction and recycling (Underwood 1988, 29).

A sustainable hazardous waste policy will also reflect the potential risks to human life and the ecosystem. Thus, priority will be given to protecting valuable resources, such as water, flora, and fauna. This can be accomplished by reducing both the volume and the toxicity of waste. Sustainable management of hazardous wastes requires that the sites of discharge must be under control and effectively monitored. The legal framework and applied regulatory mechanisms must reflect the high standards, as already practiced in some developed countries.

The concept of sustainability recognizes the need for a global scope and long-term approach, rather than a limited scope and short-term approach. We need to question our current conceptions of efficiency and equity, as they often reflect only short-term profits and national interests, rather than reaching out to bring benefits and development to all peoples.

"REALPOLITIK" OR CAN THE POLICY WORK?

A sound policy must meet another criterion: On a theoretical-political level, it has to meet the concerns of major interest groups, and on a practical-project level, it must be implementable. Changing environmental attitudes towards environmental issues is only one of those concerns, reflecting both the publics worries over worsening environmental conditions and their care for more and better environmental protection.

The discussions and negotiations on transfrontier movements of hazardous wastes reveal two major opposing positions. Industrialized countries favored free trade in wastes, arguing that a complete ban was without any chance of being adopted and that such a ban would be hard, if not impossible, to enforce. Developing countries, on the other hand, demanded a complete ban of transfrontier waste movements, emphasizing the large gap in disposal technologies and environmental infrastructure between industrial and developing countries. In their opinion, a policy allowing waste exports could not be controlled and monitored by the developing countries. Political irritations were additional important reasons to call for a complete ban. These positions show that implementation and enforcement are as instrumental as economic and political objectives in designing an effective policy.

Behind both arguments are the internal positions of political and socioeconomic nature that reflect social and economic realities in industrialized and developing countries. The most visible factor in industrialized countries is perhaps the pro-environment movement of the past decade, which has been changing governmental perceptions and politics throughout Europe. Environmental groups not only lobbied for more general environmental protection, but also for incentives to change consumption patterns to reduce wastes, for measures to improve public health, and, last but not least, for significant action to conserve natural resources. As environmental groups formed political parties and entered state and federal parliaments in European countries, governmental policies have had to adopt and integrate environmental concerns.

The various industries that generate hazardous wastes comprise another important group. These industries are the backbone of any industrial country, providing not only employment, but substantially contributing to the general welfare. While it should be in industry's own interest to manage their hazardous wastes in an environmentally sound manner, industry is concerned about the skyrocketing disposal costs. In the past, their economic power was directly used to influence governmental policy formulation and to shape major national and international policies relating to industrial development. Industry's position in this regard will change slowly, only as environmental groups with public support gain momentum and exercise pressure on all policy making levels. Given the existing power constellation, a hazardous waste policy will have to reflect stronger environmental values than ever, but also be acceptable to industry, in order to be feasible.

A future policy for exporting hazardous waste will not succeed if it is not implementable on a practical level. This would require the creation of an infrastructure, to enforce and monitor the policy, including numerous technical, regulatory, and administrative features. The detailed provisions for enforcement and monitoring, but also for other critical issues, such as public safety, liability, and insurance, will be examined for each policy in this book.

References

Ashford, N. 1981. Alternatives to cost-benefit analysis in regulatory decisions. *Annals of the New York Academy of Sciences* 363:129–137.

Brown, L.R. 1981. *Building a Sustainable Society*. Washington, D.C.: World Watch Institute.

Carr, M. 1988. *Sustainable Development:Seven Case Studies*, ed. M. Carr.

Honecker, M. 1980. Verantwortung für die Umwelt als Pflicht des Menschen. In *Vorsorge für die Umwelt*, eds. G. Hosemann, and E. Finckh. Erlangen.

Jonas, H. 1979. *Das Prinzip Verantwortung, Versuch einer Ethik für die technologische Zivilisation.*

Kelly, M.E. 1985. International regulation of transfrontier hazardous waste shipments: a new EEC directive. *Texas International Law Journal* 21(85):86–128.

OECD. 1985. *Transfrontier Movements of Hazardous Wastes: Legal and Institutional Aspects*. Paris: OECD.

Roelants du Vivier, F. 1988. *Les Vaisseaux du Poison: La Route des Déchets Toxiques*. Paris:Sang De La Terre.

Rogets II. 1980. *The New Thesaurus*. Boston.

Stivers, R.L. 1981. *The Sustainable Society: Ethics and Economic Growth.*

Tolba, M.K. 1987. *Sustainable Development, Constraints and Opportunities*. Nairobi:UNEP.

Underwood, J.D. 1988. Managing hazardous wastes is not enough. *Industry and Environment*, UNEP 11(1):29–31.

United Nations. 1973. *Report of the United Nations Conference on the Human Environment*. Stockholm, June 5–16, New York: UN.

1988. *Evolving Environmental Perceptions: From Stockholm to Nairobi*, ed. M.K. Tolba Nairobi: UNEP.

U.S. Congress. 1989. International Export of U.S. Waste. Hearing before a Subcommittee of the Government Operations, House of Representatives, 100th Congress, July 14, 1988. Washington, D.C.: Committee on Government Operations.

Weir D., and M. Schapiro. 1981. *Circle of Poison, Institute for Food and Development Policy*. San Francisco: Institute for Food and Development Policy.

Whynne, B. 1989. The toxic waste trade: international regulatory issues and options. *Third World Quarterly* 11(3):120–46.

World Bank. 1989. Economic analysis of sustainable growth and sustainable development. World Bank. Department of the Environment. *Working paper No. 15*. Washington, D.C.: World Bank.

World Commission on Environment and Development (WCED). 1987. *Our Common Future*. Oxford: Oxford University Press.

2

Hazardous Waste Exports—Recent Findings

This chapter gives an overview of hazardous waste exports—the historical development in this area, the countries of import and export, existing and planned transfer schemes, and the scope and scale of current waste export schemes. Basic information on hazardous waste management (HWM) and such definitions as exist have also been included, in order to provide a better understanding of the complex questions raised by hazardous waste exports. This chapter also examines past practices of moving hazardous waste, as an organized business, and their legal and financial implications. Finally, I will describe current developments and assess future prospects.

HISTORICAL DEVELOPMENTS

Exporting hazardous waste, as an alternative to managing it within the borders of the generating country, is not a new phenomenon of the 1980s. Already in the 1970s, some wastes were exported from France and the United States to francophone Africa.[1] The perpetrators of another export scheme, which involved the illegal export of hazardous wastes from the American armed forces to Zimbabwe by falsely labeling them as "cleaning fluids," were recently sentenced to prison for fraudulent business practices (Schissel 1988). The Seveso accident in Italy of 1976, which in itself caused severe human and environmental damage, triggered strong reactions in the European Parliament, when hazardous wastes that had not been cleaned up, including PCBs, were missing. The wastes disappeared until

---

[1]In 1987, *The Nation* reported on several early waste export schemes from the United States to India, South Korea, Nigeria, and Honduras. The U.S. Department of Commerce, which also monitors some waste export schemes, released information showing that since 1980, wastes have been exported to some 35 underdeveloped countries (Porterfield and Weir 1987).

1983, when they were found in a slaughter house in southern France. How and by whom they were transported is still unknown.

In retrospect, the increasing activity in exporting hazardous wastes relates to two developments; the tightening of U.S. laws concerning hazardous waste in the early 1980s, and the increase in attention paid by the European Community to issues surrounding hazardous wastes since the middle of the decade. In the latter half of the 1980s, the number of waste shipments and the volumes exported increased sharply. Shortly thereafter, the first cases of toxic waste dumping were found in Africa. Their discovery was occasioned by health problems of people who contracted severe headaches and other sicknesses. Although several import-export contracts of hazardous wastes involved highly placed government officials of the receiving countries, the subject did not receive significant notice. It was only after concerned journalists brought the waste dumps to public attention that the issue evolved into an international scandal. As more waste schemes were discovered, a conflict formed between industrialized and developing nations. African countries condemned the practice as "toxic terrorism" and "garbage imperialism," and subsequently took action to ban the imports of hazardous wastes. On the other hand, many industrialized countries defended the trade of wastes as a legitimate business practice already well-established between industrialized countries, and which benefits both parties. The differences between the countries in the Northern and Southern hemispheres seemed insurmountable and even strained diplomatic relations between some countries.

In order to understand some of the implications of transfrontier movements of hazardous wastes, an important distinction needs to be made between transferring hazardous wastes among industrialized countries and transferring from industrialized to developing countries. Transferring wastes among industrialized countries can generate mutual benefits, given their resources and management capability. With proper technology and effective environmental regulation, the wastes can be recycled or managed, according to state-of-the-art techniques. So far, there is no evidence that the trade of waste within OECD countries has led to environmental problems. This is not the case in developing countries, where many past waste imports were dumped improperly, often without control or monitoring. The results were direct hazards to human lives and damage to the ecosystem.

National governments throughout the world, as well as regional and international organizations, have since addressed the issues of hazardous waste exports in their policy-making bodies. A variety of laws have been passed and an international convention to control transfrontier movements of hazardous waste was signed in 1989. Two alternative policy directions have evolved from all these activities: (1) to try to effectively control and monitor transboundary movements of hazardous wastes, and (2) to ban all movements of hazardous wastes. An analysis of these policies is provided in Part 2.

DEFINITIONS AND CLASSIFICATION

Assessing the scope and scale of hazardous waste exports requires a look at the definitions and classifications of hazardous wastes. Currently, the systems of defining and classifying hazardous wastes vary widely among different countries. In general, there is a distinction between a *definition* of waste and a *system for classifying* materials that have been defined as a waste (CEFIC 1989). Following is a brief overview of the definitions and classification systems primarily used in industrial countries, which produce by far the largest amounts of hazardous wastes.

Using various definitions for wastes, such as "hazardous," "toxic," "dangerous," and "special," is the cause of many misunderstandings and a large impediment for forming a global system of hazardous waste management. Therefore, this chapter will not detail specific national definitions and systems of classification[2].

There are several justifications for the call for a uniform international legal definition and classification system for hazardous wastes. First and foremost, a common legal definition is necessary for resolving disputes that arise from exporting and managing hazardous wastes. Furthermore, clear definitions would also distinguish hazardous wastes from other materials appropriate for recycling, reclamation, reuse, and resource recovery, and, moreover, provide a necessary link between civil law and environmental law.[3] Presently, civil law controls the exchange of commodities that have a positive value to the transactors by transferring title, property, and risk. This is in contrast to environmental law, whereby a hazardous waste generator may not contractually transfer the risks (i.e., liability) to another party (Yakowitz 1989, 5-6). A clear definition, with its corresponding classification system, would clarify which law is applicable and also assist authorities in implementing and enforcing legislation, to monitor and control hazardous waste.

Studying the definitions used for hazardous waste in various countries indicates

---

[2]For further information on this important issue, refer to Harvey Yakowitz, who compared and analyzed the systems from selected countries (Yakowitz 1988, 7-8).

[3]The definition of hazardous wastes subject to transboundary movements generally *excludes* radioactive wastes, and neither the OECD policy nor the Basel Convention on the Control of Transboundary Movements of Hazardous Wastes and their Disposal included radioactive materials or wastes. Radioactive materials, including wastes, are part of other international negotiations convened by the International Atomic Energy Agency (IAEA). The IAEA general conference in 1988 has established a "technical working group of experts to develop an internationally agreed upon code governing nuclear waste trades." According to Greenpeace International, regulations for radioactive wastes—excluding fissile radioactive wastes (capable of producing a fission reaction)—do not exist, and it is currently possible to export and dump radioactive wastes, without international control. Greenpeace further points out that a Code of Practice for radioactive waste is not a binding legal mechanism and has therefore called for the inclusion of radioactive wastes in ongoing negotiations (Greenpeace International 1989a).

that no two systems are alike, and some are even inconsistent with each other, which creates numerous problems (Hannequart 1985). An internationally acceptable legal definition would reconcile the different national definitions and assist, as a precondition, the development of universal standards for hazardous waste management. This appears to be even more important for waste recycling. As the European Community approaches a single, unified market, waste recycling will increase, as any other trade would. In order to guarantee equally high safety and environmental protection in all member states, definitions and other environmental standards have to be harmonized.

In order to achieve such an international system, the most important group initiative came from the OECD, whose members comprise 24 industrialized countries. The OECD Council adopted, in 1988, the following definition:

1. "Wastes" are materials, other than radioactive materials, intended for disposal for specified reasons.
2. "Disposal" means any of the operations specified as disposal operation.
3. "Hazardous wastes" are defined as (1) a core list of 44 wastes, and (2) as "all other wastes which are considered to be or are legally defined as hazardous wastes in the Member country from which these wastes are exported or in the Member country into which these wastes are imported" (OECD 1988).[4]

Defining wastes is the first step in developing a system of classification for wastes. Several international organizations have been working on the promulgation of classification systems for specific purposes: the European Community established, in 1978, a list of toxic or dangerous substances and materials, which was based on an EEC Council Directive dealing with toxic and dangerous wastes; the United Nations Committee of Experts on Transport of Dangerous Goods promulgated the UN Recommended Classification System (UN-RCS); and other lists that have been promulgated through international sea dumping conventions (Yakowitz 1985). However useful these systems are for their specific purposes, none of them provides a complete framework for controlling transboundary movements of hazardous wastes.

The OECD approach to defining wastes also includes a classification system, which is the most comprehensive to date. A series of seven tables is used to define and classify hazardous wastes to be controlled when subject to transfrontier movements. These tables are comprised of:

1. A "core list" of wastes to be controlled;
2. Reasons why materials are intended for disposal;

---

[4]This core list of 44 wastes is also the basic list of hazardous wastes of the Basel Convention.

3. Disposal operations;
4. Generic types of potentially hazardous wastes;
5. Constituents of potentially hazardous wastes;
6. List of hazardous characteristics; and
7. Activities that may generate potentially hazardous wastes (OECD 1988, 3-4).

Each of these lists carries a specific letter and number, and the combination of letters and numbers, which refer to the above listed criteria, provide a complete reference system (Yakowitz 1988, 3-10). As such, it represents the International Waste Identification Code (IWIC) adopted by the OECD, and by UNEP in the Basel Convention.

This system allows a fairly practical application and is simple to interpret, provided the correct translation is made into the various languages. For developing countries, which might not yet have a system, it could be of guidance to establish a national list that meets their particular environmental, economic, and sociopolitical conditions. In addition, this system seems to be flexible enough to allow changes or additions, as new hazardous chemicals emerge and new recycling technologies become available.

IMPORTING AND EXPORTING COUNTRIES

As the previous discussion shows, applying various definitions to and distinguishing between *wastes* and *hazardous waste* shipments is difficult to decipher, and there is no clear dividing line. Both industrialized and developing countries import hazardous wastes. In industrialized countries, these are either "managed" in form of recycling, reuse, or incineration, or disposed of in landfills. In developing countries, managing imported hazardous wastes most often consists only of their final disposal in landfills. Historically, the trade in wastes was conducted within industrialized Europe, and between the European Community and North America.

Currently, about 90 percent of the requests for hazardous waste shipments from the United States abroad are to Canada, and the rest is mainly to Mexico and other countries (Hansen 1989). From exported Western European wastes, 80 percent are sent to other Western European countries, 15 percent to Eastern Europe, and 5 percent to the developing countries, according to the OECD (Greenhouse 1989). Before the revolutionary political changes in Eastern Europe, most hazardous waste exports from Western Europe to Eastern Europe were from Germany (West Germany) to the former German Democratic Republic (East Germany). Since the unification of these two countries, exporting hazardous wastes—but not household wastes—has been terminated, whereas export to Poland and other Eastern block countries has increased (see "Eastern Europe," in Chapter 5).

TABLE 2-1.   Countries that Received Wastes Between 1986 and 1988, by Region.

Africa	The Americas	Asia/Middle East	Europe	South Pacific
Guinea	Canada	Japan	Finland	Australia
Nigeria	Brazil	Lebanon	Netherlands	Papa New Guinea
S. Africa	Haiti		Sweden	Solomon Isl.
Zimbabwe	Mexico		U. Kingdom	
Benin	Argentina		W. Germany	
Gabon	Dominican Rep.		Belgium	
S. Leone	Guyana		Italy	
E. Guinea	Panama			
	Paraguay			
	Peru			
	Surinam			
	Uruguay			

Source: Greenpeace International 1989, The International Trade in Waste: A Greenpeace Inventory

The number of countries that, in the past, imported, intended to import, or were proposed for imports of hazardous wastes increased sharply in the 1980s. Since most of the hazardous waste exports to nonindustrialized countries were covert operations, often including sham-recycling and criminal activities, it is very difficult to identify all the waste schemes and countries involved. From the available information and analysis conducted by the EPA, the National Investigation and Enforcement Center of the EPA, and the Hearing of the U.S. Congress on this subject, it can be assumed that many of the waste shipments prior to 1986 were known neither to the public nor to the authorities.[5]

An inventory of past and present waste schemes was conducted by Greenpeace International which formed the International Toxic Waste Action Network, in order to document and investigate waste export schemes (Vallette 1989). Due to its international network, the Greenpeace inventory can be considered the most comprehensive documentation of waste export schemes to date. According to Greenpeace data, at least 11 developing nations received wastes between 1986 and 1988, namely Brazil, Guinea, Haiti, Lebanon, Mexico, Nigeria, Sierra Leone, South Africa, Syria, Venezuela, and Zimbabwe. Another 38 developing countries have been proposed for hazardous waste imports. Table 2-1 lists countries that received wastes between 1986 and 1988, or that are expected to receive wastes under "active proposals that have not yet been implemented." Table 2-2 lists all countries, except Eastern European countries, that have been proposed for hazardous wastes disposal (Greenpeace International 1989, 386). From the tables, it can

---

[5]The Inspector-General of the U.S. EPA testified that the EPA does not know whether it controls 10 or 90 percent of the wastes shipped abroad. See Chapter 6 for an in-depth analysis of U.S. policy.

TABLE 2-2.   Countries that Have been Proposed for Hazardous Waste Disposal, but Have not Accepted These Wastes.

Africa	The Americas	Asia/Middle East	Europe	South Pacific
Benin	Argentina	South Korea	Belgium	Philippines
Congo	Bahamas	Syria	France	Papua N. G.
Gabon	Belize	Taiwan	Spain	Solomon Isl.
Gambia	Bermuda	Pakistan	Sweden	Tonga
G. Bissau	Costa Rica		Greece	W. Samoa
Marocco	D. Republic		Netherlands	Samoa
Djibouti	Guatemala			Hong Kong
Guinea	Guyana			
Nigeria	Honduras			
Senegal	Martinique			
	N. Antilles			
	Panama			
	Paraguay			
	Peru			
	Uruguay			
	Chile			
	Haiti			
	Mexico			
	Surinam			
	Venezuela			
	Turks & Caicos Islands			

Source: Greenpeace statement on the *International Export of U.S. Waste. Hearing before a Subcommittee of the Government Operations,* House of Representatives, 100th Congress, July 14, 1988. Washington, D.C.: Committee on Government Operations.

be concluded that, aside from the substantial trade among industrialized countries, most waste shipments went to the Americas and to Africa.

Two other studies (Roelants du Vivier 1988, 90–91; Centre Europe-Tiers Monde 1989), which mainly concentrate on the African continent, also document waste export schemes. Figure 2-1 shows that almost all West African states and a number of other African countries were targeted for hazardous waste schemes. Many agreements and contracts have been published to prove the existence of this trade. The lists speak for themselves: a large number of states are involved in a trade that is unmistakably global. Any coherent policy in this field must therefore be concerned with an international scope.

GENERATION AND EXPORT

Managing hazardous wastes undoubtedly represents an increasingly important problem, as the volumes of wastes generated increase throughout the world. It is

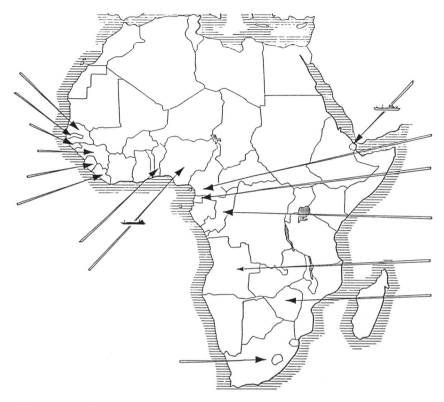

FIGURE 2-1.   Western African States Being Proposed for Hazardous Waste Imports.   Source: Bureau de Reportage et de Recherche d'Informations (BRRI). 1989. Afrika hungret, Da habt ihr unsern Dreck!, Lausanne, Schweiz.

important to note that the data available on hazardous waste generation are based on the *differing* definitions applied in industrialized countries. Some sources may also not be as reliable as others and therefore all composites contain partially contradictory figures. Furthermore, in addition to the problems caused by different definitions and doubts about the reliability of sources, uncertainties also arise from the variations in the scope of the estimates, in the measures and the sampling, and from response errors. As a result, the data available needs to be evaluated very carefully. However, in order to assess the scale of the hazardous waste export schemes, and to decide whether a global policy is necessary, data on the generation of hazardous wastes and their exports are required. Before we can look at exports, an estimate of the global generation of hazardous wastes is helpful.

The amount of wastes generated annually, and that can be considered hazardous, is large. Presently, the EEC produces a total of about 2,200 million tons of

TABLE 2-3.   Production of Hazardous Wastes in the European Community.

Country	Quantities of Hazardous Wastes as Available, in (1000 Tons/Year)	Population	GNP (Million ECU)	Estimate Calculated According to GNP (1000 Tons/Year)
Belgium	1000	9,860,000	69,657	700
Denmark	67	5,100,000	52,355	500
Spain	1800	38,370,000	144,662	1500
France	4300	55,300,000	397,127	4000
Greece	—	10,000,000	32,316	300
Ireland	76	3,630,000	14,191	100
Italy	2000–5000	57,320,000	282,998	2800
Luxemburg	—	367,000	3,888	50
Netherland	1000	14,520,000	112,093	1100
Portugal	—	10,100,000	18,224	200
W. Germany	4500	61,010,000	511,573	5000
United Kingdom	3700	56,400,000	382,244	3750
Total EEC	18,443–23,443	321,977,000	2,021,328	20,000

Source: Mettelet, C. 1989. Production, Traitement, Recyclage et Transferts Transfrontaliers de Dechets Dangereux dans la Communite Europeenne. Agence Nationale pour la Recuperation et L'Elimination des Dechets—Les Transformeurs. Paper read at the workshop on Hazardous Waste Management Beyond 1992, 25./26. April 1989. Scientific & Technological Options Assessment (STOA), Europena Parliament, Brussels. (Data: Estimation by ANDRED—Les Transformeurs, France 1989.)

wastes, of which an estimated 20 to 30 million are hazardous (Commission of the EC 1989, 3; Mettelet 1989, 3). Within the European Community, data on the generation of wastes and hazardous wastes exist for almost all countries. Based on these data, one can calculate an EEC-wide average rate of hazardous waste generation of about 10,000 tons/year for each billion ECU[6] of Gross National Product (GNP) (Mettelet 1989, 3), despite the substantial variation in population and economic strength of EEC member countries. These calculated tons are shown in Table 2-3. Column 2 lists the figures of hazardous waste generation, as they are available for certain countries; column 5 gives, for comparison, the calculated figures based on GNP. For the other OECD countries, the OECD Environmental data compendium for 1989 provides a set of figures, which are listed in Table 2-4. The OECD data are also compiled from sources using different definitions—*hazardous, industrial, special*—and therefore cannot be considered infallible.

Altogether, when one adds estimates of the amounts generated by the Eastern European countries, the newly industrialized countries (such as Malaysia, India, Thailand, Indonesia, and so on), and as the developing countries (which in total are estimated to be 5 million tons or more) (Roelants du Vivier 1988), to those of the

---

[6]1 European Currency Unit (ECU = U.S. $1.35 (January 1991)

TABLE 2-4.   Hazardous Waste Generation in the OECD.

Country	Year	Hazardous and Special Wastes in 1000 Tons
Canada	1980	3,290 a)
United States	1985	265,000
Japan	1985	666 b)
Australia	1980	300
European Communities	c)	17,721
New Zealand, Austria, Finland, Norway, Switzerland	d)	609
OECD	TOTAL	287,586

a) Wet weight; b) 1986; c) Different years from 1980–1987.

Source: OECD. 1989. *OECD Environmental Data Compendium 1989*. Paris: OECD.

industrialized world, one finds that between 300 to 500 million tons of hazardous wastes are generated each year worldwide (Roelants du Vivier 1988, 75).

UNEP and the OECD estimate that the quantity of waste managed abroad, and thus subject to transfrontier movement, between 10 and 20 percent, which means some 30 million or more tons of waste movements per year. One analysis maintains that approximately 80 percent of these wastes are exported within OECD countries, 10 to 15 percent are exported to Eastern Europe, and 5 to 10 percent are exported to developing countries (Roelants du Vivier 1988, 75). However, official OECD export and import data cannot fully account for these amounts.

The OECD includes, in its latest environmental data report, import and export figures concerning the total generation of hazardous wastes in each member country. A representative sampling of these figures is given in Table 2-5, which provides data on hazardous waste imports and exports for selected countries. From this table, it can be concluded that the volume of hazardous wastes that are exported from the industrialized countries, but not imported to other OECD countries, is substantial. The figure is obtained by subtracting the total volume of hazardous waste imports from the total volume of hazardous waste exports. The calculation gives 654,000 tons, or 54.6 percent of official exports, which is almost certainly sent to Eastern Europe or developing countries. It would thus appear that more than half of the officially acknowledged volume of exported waste is not sent to industrialized countries, but to less developed countries. Furthermore, according to other sources, West African countries imported up to 24 million tons of hazardous wastes in 1988 alone (International Environment Reporter 1989, 50), and the OECD listed 500,000 tons of hazardous waste imports to East Germany from non–OECD-Europe in 1983 (OECD 1985, 50).

One also sees from the table the very heavy reliance of some countries (most notably Ireland, Luxembourg, and Switzerland) on exports, to dispose of their

TABLE 2-5.   Hazardous Waste Import and Export of Selected Countries.

Country	Year	Import 1000 Tons		Export 1000 Tons	
Canada	1980	120	3.6%	40	1.2%
United States	1987	40	—	150	0.1%
Denmark	1985	—	—	20	16%
France	1987	250	12.5%	25	1.3%
Germany (F.R.G.)	1985	50–100	1–2%	700	14%
Ireland	1984	—	—	20	100%
Luxembourg	1985	—	—	4	100%
Netherlands	1986	—	—	155	10.3%
Sweden	1980	—	—	15	3%
Switzerland	1987	—	—	68	56.7%
United Kingdom	1986	83	2.1%	—	—
Total		543		1197	

Source: OECD. 1989. *OECD Environmental Data Compendium 1989.* Paris: OECD.

hazardous wastes. For the discussion that follows in Chapters 3 and 4, on the reasons and consequences of hazardous waste exports, some additional figures compiled by United Nations sources are presented below. Despite the limited quantity of information available in the United Nations, the figures are useful, to illustrate the magnitude of the issue:

1. Canada exported 70,000 tons in 1982, compared to 40,000 in 1980 (United Nations 1987).
2. France imported 73,200 tons in 1976, compared to no imports in 1987 (United Nations 1987).
3. Hungary is reported to be generating 7,081,000 tons of hazardous wastes (United Nations 1987).
4. The United States has exported to Guinea, South Africa, Zimbabwe, Haiti, Mexico, Brazil, and Canada (UNEP 1989).
5. "Italy is known to export large quantities of waste to Africa and the Middle East but no data are available" (UNEP 1989, 8.11).
6. Japan generated 292 million tons of hazardous "industrial chemical waste" in 1980 (UNEP 1989), compared to the less than 1 million tons reported by the OECD.
7. India generated 35.72 million tons (1980) and Malaysia generated 418 million m^3(1985) of hazardous wastes per year (UNEP 1989).

The last two items show that attention should be paid to hazardous waste movements in Asia and Southeast Asia, where not only does little data exist, but also where virtually no hazardous waste research has been conducted to date. With increasing industrial activities, particularly in the electronic industry, which is booming in some of these countries, the rising volumes of highly toxic wastes pose a serious problem.

The case of Japan provides an interesting, and hardly unique, illustration of the problems in the region. Japan had its own tragic episodes with the Minamata Disease, the Itai-itai Disease, and so on, which resulted from toxic waste dumping and other unprotected hazardous occupational activities in the past. Japan's current regulation of hazardous wastes is basically a result of these diseases, as "seldom has been the case when chemicals became being [sic] regulated for preventive measures" (Gotoh and Okazawa 1989, 116). According to UNEP, Japan generated 292 million tons of hazardous "industrial chemical waste" in 1980 (UNEP 1989), and 312.3 million tons of industrial wastes in 1983 (Ministry of Health and Welfare, Tokyo 1990, 18). A Japanese study concludes that "compared to North America and Europe, the disposal cycle of waste is faster in Japan. That is, consumer durables are used for shorter periods and development of disposable products is more frenetic" (Ueta 1990).

Faced with an increase in the volume of waste, and with spiraling waste management costs and facility siting problems caused largely by public opposition, the government undertook various measures to increase disposal capacity and to support the recycling of wastes (Fujiwara and Tanaka 1989). However, the limited capacity for hazardous waste management still requires that some wastes be shipped to areas further away from industrial centers. For example, it is reported that some hazardous wastes are sent from Chiba to Aomori, and from the Kansai area to islands in the Inland Sea (Ueta 1990). Furthermore, due to the long distances to disposal sites and increasing costs of disposal, "illegal disposal of industrial wastes…is quite common" and an "economic phenomenon" (Ueta 1990). And as recently as 1989, serious illegal waste dumping schemes involving 8000 drums of highly toxic wastes (PCBs) were arranged in Japan (Ueta 1990).

Two further problems can be identified in Japan. Waste materials that were once deemed valuable for recycling or resource recovery are no longer considered waste. Therefore, some of the hazardous wastes intended for recycling are not controlled or regulated (Gotoh, Okazawa 1989). Furthermore, the latest 5-Year National Program for the Construction of Waste Treatment Facilities (1986-1990), for which the government proposed U.S. $13.6 billion and whose objective is to build waste disposal facilities throughout the country, includes constructing large-scale regional waste landfilling sites, as "sea reclamation" projects. These projects involve the dumping of (hazardous?) wastes on the shore, in order to gain additional land.

Unmistakably, all figures have to be interpreted very cautiously. Given the different definitions and the lack of a uniform and comprehensive classification system (the result of which most likely is lower official estimates), expert assessments that available figures on import and export of hazardous wastes present only "the tip of an iceberg" may be very close to the truth. Even the officially reported trade in hazardous wastes is enormous: The 1985 OECD-report stated that approximately 100,000 transfrontier movements, covering about 2.2 million tons of hazardous wastes, occur each year, which means that "in the OECD, a cargo of hazardous wastes crosses a national frontier more than once every five minutes, 24 hours a day, 365 days per year" (OECD 1985, 7).

Notwithstanding the range of figures for worldwide hazardous wastes exports (from millions to possibly tens of millions of tons per year), the potential harm of the officially acknowledged 654,000 tons of waste not exported to industrialized countries and possibly dumped in other countries underlines the need for a global policy, to control and manage these wastes in an environmentally sound manner.

AN ORGANIZED BUSINESS

The business of waste management in the country of origin and its export to other countries is a multifaceted enterprise, involving many different actors, strategies, circumstances, and activities.

Actors

With increased generation of hazardous wastes, waste management has grown rapidly to a US$ 4 billion industry in the United States alone (Scientific American 1989, 89). In the European Community, the turnover in the waste business sector is estimated to be up to 200 billion ECU (Commission of the EC 1989, 4). Some waste management firms also have expanded their activities abroad, and more and more transnational corporations are becoming involved. Many of these firms carry out a decent and straightforward activity. However, an increasing number of the companies in the business are made of "discreet brokers and intermediaries, complaisant shipping firms, and sundry ghost companies," (Schissel 1988, 48) registered in rather unusual places, such as the Isle of Man or Gibraltar. In the waste export schemes of the past few years to West Africa, it has become rather difficult to identify the individuals behind the companies involved. One example is the Hobday Company. The Isle of Man company register lists the directors of Hobday as Martin John Gibbs, a building contractor, and his wife, Margaret, a teacher, with an address in the village of Eremi, near Limassol in Cyprus. Enquiries in Cyprus, however, have revealed that the Gibbses sold their house and moved to Gibraltar" (George 1988, 39). Another figure entering the waste trade business was Andreas Künzler, a Swiss businessman who is known as an "arms dealer" and "mercenary" in several African countries. Furthermore, many of these new waste brokers are incompetent in hazardous waste management. As an EPA official states, "...many brokers don't

know how to handle the material they want to ship. They are not technical people...with no expertise in biology or toxicology" (Sheppard 1988). It can be
concluded that new firms attracted by this business, primarily by the high profits,
often lack the capacity or the willingness to perform according to regulations.

Motivation

The main reason why more and more individuals and firms enter the waste trade
business is because of the huge profits that can be made in brokering hazardous
wastes. Current treatment of hazardous wastes, such as PCBs, costs up to US$
3,000 per ton. Exporting hazardous wastes is much cheaper. As many contracts
reveal, the payments are as little as US$ 2.50 per ton. Even with the additional
transportation costs, this is an enormous price differential, which culminates in
significant profits (see also Case 3) (Schissel 1988, 48). One result of the high
profits is that the actors involved may begin falsifying documents or showing other
signs of incipient corruption, in order to preserve their gains. Corrupt individuals
were involved in many waste schemes, in which the governments were not
informed at all. The waste schemes were thus private business transactions,
without any authorizations, permits, or oversight.

The incentives offered to developing nations present another risk. In addition to
bribing individuals, smaller or larger "presents" for debt-ridden countries are
offered. In cases where governments were involved, some waste schemes were
signed at the ministerial level. Often, such contracts were kept secret, and some of
the documents now available for study are still marked "confidential." Furthermore, the profits involved are not only of the financial kind. In one case, an Italian
broker managed the export of 2,500 tons of hazardous wastes, to be disposed of in
Lebanon for a payment that is rather uncommon—the exchange of weapons (Grefe
1988, 8).

Strategies, Circumstances, and Activities

Basically, there are three kinds of activities in the export of hazardous wastes: (1)
compliant waste handlers, (2) sham recyclers, and (3) criminal activities. All these
entail their own particular risks and problems.

Compliant Waste Handlers

According to the EPA's Office of Enforcement, most of the facilities known to
export or import hazardous wastes can be considered as "wishing to comply with all
applicable regulations" (National Enforcement Investigation Center 1988, 13).
Sometimes, however, wastes are exported but are not defined as "hazardous," such
as incinerator ash or polluted soil. Often, these are loaded with lead, mercury, and

other toxic materials. In such a case, the waste export scheme is not regulated at all, but may nevertheless cause serious difficulties, as Case 1 of the "Khian Sea" shows.

CASE NO. 1   The Khian Sea with "nonhazardous" waste.

The 1988 hearing before the U.S. Congress revealed the details of one of the most serious waste export deals.

"In August 1986, the Khian Sea left the port of Philadelphia bound for the Bahamas, with 15,000 tons of municipal incinerator ash. However, the ship was turned away from there with its cargo. Thus began a rather unbelievable odyssey which has lasted almost 2 years. It involves a number of countries…and at times has caused some diplomatic problems. It made numerous stops including Fort Lauderdale, Puerto Rico, the Antilles, the Dominican Republic, Jamaica, Panama and the Cayman Islands, before dumping 2,000 tons of ash on the shores of Haiti in February (1988)" (Synar 1989). In 1986, its cargo was described as "non-hazardous, non-flammable, non-toxic incinerator ash." When the ship left Fort Lauderdale in March 1987, the Shippers Export Declaration called the cargo "general cargo." Then in December 1987, when it left Caymen Crac, the cargo was described as "bulk construction material," and when it was finally dumped in Haiti, it was described as "top soil ash fertilizer" (Synar 1989, 112). The cargo was still the incinerator ash. In January 1988, the Khian Sea left Haiti, leaving behind 2,000 tons of incinerator ash and causing a "quite strong reaction" in Haiti, which drew international attention. Both the Pan-American Health Organization (PAHO) and the World Health Organization (WHO) were involved in the aftermath of that waste dump. Several independent studies have recommended the "removal" of the ash (Synar 1989, 126). This episode is still far from being resolved.

In essence, a ship loaded with 15,000 tons of incineration ash was constrained to sail for two years from port to port before finally dumping its unwanted cargo in Haiti, where it is still, two years later, the subject of considerable debate. The papers accompanying the ship were falsified, and the receiving country, the Bahamas, was not provided with adequate information about the cargo. (In the United States, as in many other countries, a requirement for a manifest[7] does not exist for nonhazardous wastes). As a result, the importing country could not make an informed decision. In retrospect, it also appears that there was no concern over where the waste would end up. And given the current U.S. regulations, this incinerator ash was considered nonhazardous and thus was not controlled by the EPA.

---

[7] A "manifest," "trip-ticket," or "consignment note" are shipping papers that give all necessary information and data about the generator, transporter, and disposer and about the cargo, including a description of the nature, contents, and hazards of the wastes being shipped.

Sham Recycling

Another form of waste export is "sham recycling." This activity is increasing, according to EPA. information compiled from export notices and inspections (NEIC 1988, 14). Sham recycling occurs when materials have been classified as *recyclable* and are destined for recycling, but when the actual management process involves disposal. Another form of sham recycling is when hazardous wastes are misclassified, in order to avoid appropriate regulatory control on the generation, transport, and disposal of the wastes. Misclassification can arise from a lack of information, a misunderstanding of regulations, ignorance, or from a cognizant attempt to avoid regulation. At the U.S. congressional hearing in July 1988, the Inspector-General of the EPA reported that an internal investigation "found instances where hundreds of tons of hazardous waste were exported without notifications of intent to export filed with EPA." (Martin 1989, 19). This included ten shipments, with an accumulated weight of almost 1,000 tons. Let us look at such an example in Case 2.

CASE NO. 2.   Turkey: Toxic waste as "economic good".

The plan was to export to 50,000 tons of sludge from Daimler (Mercedes Benz), Bosch, and Siemens to Turkey in 1987. In a trial period, 1,581 tons of this "highly toxic" waste, which were not permitted to be incinerated in West Germany, were exported to a cement plant in Turkey. The industrial wastes, consisting of more than 168 substances, were labeled as "fuel substitute" and were delivered for burning to a Turkish cement firm. Among the substances were not only lead, chrome, and copper, but also cyanide, di-chloromethane, and dichloroethylene, which can result in the SEVESO dioxene, if improperly incinerated. Without knowing the substances in the wastes, the Chamber of Commerce of Isparta informed the German authorities that their cement plant can burn "any kind of fuel." At that time, the head of the Chamber of Commerce was also the owner of the cement plant (Öko-Institut 1988, 31).

The German authorities knew about the transfer of the waste and its content, but granted an export permit for the total amount. Although the results of the chemical analysis were known to the authority, the shipment was legal. The wastes were mixed with sawdust and consecutively relabeled as a "fuel substitute," which was then considered an "economic good," not covered by hazardous wastes regulations. The waste was shipped to Turkey.

A chemical analysis by the University of Ankara, Turkey, revealed "high" levels of polychlorinated byphenyls (PBCs), upon which Turkish authorities ordered the return of the wastes. The German firm refused to import the wastes and, only after a meeting between the ministers of Environment from West Germany and Turkey, the German government accepted that the wastes were "labeled false" and agreed to transport the wastes back (Erzeren 1988).

This incident was later brought before the German parliament, in which the government stated that according to German law, the export was not illegal. By then, the wastes were still in Turkey, and a binational working group was set up, to negotiate the modalities for the reimport (Deutscher Bundestag 1988, 3–5).

This project has caused a variety of problems for both countries involved and raises serious questions, not only about the implications of present legislation regulating hazardous wastes and their exports, but also about the role of the authorities. Why was the information about the substances in the waste not transferred to the Turkish authorities? What was the role of the Chamber of Commerce and its head? Why did the control mechanism fail when the wastes were mixed with other materials?

Sham recycling has been one of the most used tactics in exporting hazardous wastes. Often, crucial information about the content, hazards, and risks of the wastes to be exported are withheld from the country of import. In addition, the materials are described as *economic goods*, such as "road paving material," "top soil fertilizer," "construction material," or "fuel substitute."

With reference to "economic goods," it should be noted that certain hazardous wastes are recyclable materials and thus could represent some economic value to other countries. Some precious metals, for example, may be very valuable to countries that do not possess these resources. Recycling might be an economically beneficial activity. If recycling makes economic sense, the wastes should not be exported. In the Turkey case, the burning of the waste oil was prohibited in West Germany and enforced regulations would have prevented its burning. In Turkey, however, the burning was allowed ("any kind of fuel" can be burned), and the project was stopped only because of strong public opposition in Turkey and West Germany.

Criminal Activities

In many countries, the existing regulations concerning hazardous waste exports are inconsistently enforced or seem to be unenforceable. In the case of the United States, and EPA official stated that "many exporters don't even bother to give notice to the agency because there isn't any enforcement" (Porterfield, Weir 1987, 341). Comparing notices to the EPA for hazardous wastes exports and the customs service records at the U.S. ports shows that "many more" shipments occurred than were actually reported to the EPA, and that the amount of trade in hazardous wastes could be as much as "8 times more" (Porterfield and Weir 1987, 341).

The loopholes in national and international legislation, as far as legislation exists at all, are numerous. Basically, they are found at the country of export, as well as the country of import, in which case both countries are often involved (see Case 3). Regarding the notification procedures, an EPA official described the process of hazardous waste exports as follows: "We have companies with all kinds of import permits and certificates from ministers of (importing) countries but when we sent notification statements the shipments were rejected" (Sheppard 1988). Likewise, some countries require waste exports to be accompanied by a manifest. However, in reality, the manifest often appears to be very different from the original design and purpose. An official description of hazardous wastes that were "inflammable, highly toxic and reactive" read as follows: "The entire load is made up of reused barrels...their volume, weight and contents are unknown. Some of them are quite damaged and have been wrapped in plastic bags such that any possible leaks cannot be identified. The barrels...are for the most part damaged and punctured...." (Roelants du Vivier 1988, 50). There was no relevant information on toxicity, and so forth, and the description certainly does not meet the minimum of required information needed to make an informed decision about sound management of the wastes, nor about the technical capability for its import (see also Case 3).

CASE NO. 3.   Toxic Wastes to Nigeria.

"The discovery was made on June 2, (1988), after a letter was received by Nigerian students in Pisa, Italy. The students, alerted by reports in local Italian newspapers, wrote to their government warning that waste from Italy was being dumped in Nigeria" (Vir 1989, 23).

An official inquiry found that the wastes were brought to Nigeria in five shipments between 27 August 1987 and 19 May 1988. Although Nigerian authorities had given permission to a Nigerian firm to import wastes for "reprocessing," some of the drums were falsely labeled so that they would not be recognized as toxic wastes (Ogunseitan 1988, 15). An Italian businessman, Gianfranco Raffaelli, persuaded a retired Nigerian timber worker, Sunday nana Nana, to store thousands of drums of wastes for U.S.$ 100 a month in his backyard, near the river port of Koko. In collaboration with a Nigerian firm, Iruekpen Construction Company, Raffaelli "falsified and forged documents and permits" for importing the drums. After additional "gifts" to Nigerian health inspectors and customs men, he brought 3,884 tons of wastes into the country, making a profit of U.S$ 4.3 million (*U.S. News & World Report* 1988, 55).

Reports say "about 4,000 of the drums were old and rusted and some were swelling because of the heat. Scientists have determined that many of the drums contain volatile solvents and there is a risk of fire or explosion that would produce highly toxic smoke....Chemical name markings include 'Polychlorodifenile' (PCBs), 'fluorosilicate', 'erocitus', and 'rheoe 53'." (Vir 1989, 23). Among the wastes, methyl melamine waste originated from Norway, dimethyl and ethylacetate formaldehyde came from Italian chemical manufacturers, and PCBs from a Turin-based electromechanical plant. "The experts' reports have conflicted to some degree, but the presence of highly toxic PCBs has been confirmed" (Vir 1989, 23).

Since the discovery of the wastes, there have been reports of "premature births and 19 deaths from contaminated rice" (*U.S. News & World Report* 1988, 55). On July 12, 1988, after the Nigerian government ordered the waste to be sent back to Italy, dock workers began to repackage the waste into containers when three workers suffered severe chemical burns while moving the drums. "Doctors at the site reported that some of the crew were vomiting blood...and one man had been partially paralyzed" (Vir 1989, 23).

Following these disclosures the Nigerian government recalled Nigeria's ambassador from Rome and ordered the seizure of the Danish ship *Danix*, which transported the wastes to Nigeria. In Nigeria, 40 people were arrested for "conspiring to bring the waste into the country, and the government announced that the death penalty could be imposed for anyone found guilty of illegal dumping of toxic waste" (*International Environment Reporter* 1988a, 375). On June 10, 1988, the Italian merchant ship M.V. Piave was seized, to transport the wastes back to Italy.

After lengthy and heated negotiations between the Nigerian and Italian governments, the leaky drums, barrels, transport containers, and bags of chemicals from the United States and ten European countries were loaded on the *Karin B.*, a West German ship, which was to bring the waste back to Italy. The first Italian harbor, Ravenna, rejected the load because of "extremely violent opposition from the civilian population and environmentalists" (*International Environment Reporter* 1988b, 470). En route with its freight, the *Karin B.* was also refused entry in Cadiz, Spain, and the United Kingdom banned the ship from British ports. The French government subsequently prevented the ship from entering into French territorial waters and placed the vessel under military surveillance. Finally, months later, the *Karin B.* and another ship, the *M.V. Deep Sea Carrier* returned the wastes to Italy.

In cases where a sophisticated legal framework exists (e.g., in the bilateral agreement between the United States and Mexico), a "lack of coordinated enforcement strategies between the two countries and inadequate funding to carry out a dual enforcement program are common pitfalls" (Exchange 1988, 6). And auditors of the EPA Program to Control Exports of Hazardous Wastes concluded that "brokers in the U.S. could disregard the regulations for hazardous wastes exports with slim chance of being detected" (*U.S. News & World Report* 1988, 55). Furthermore, U.S. customs officials were not trained to sniff out illegal shipments (*U.S. New & World Report* 1988), as funding is inadequate. The government is powerless to stop shipments that it knows are dangerous or to advise countries against accepting them because "we have to transmit notification with no value judgements at all" (*U.S. News & World Report* 1988, 55).

PRESENT DEVELOPMENTS

In the past several decades, the volumes of household and hazardous waste generation have increased, partially exponentially. In the early 1980s, the EEC generated more than 2 billion tons each year, and the volumes in these 12 member states are increasing by 70 million tons/year. The share of wastes that are hazardous will increase correspondingly.

In a parallel phenomenon, the transboundary movement of wastes has also increased. The United Kingdom is one of the few industrial countries that import hazardous wastes on a large scale. The amounts imported to the United Kingdom have increased steadily over the last decade. Other available data show even higher imports. The increase has occurred in the number of shipments, the volumes transported, and the number of countries involved (Benjamin 1989).

In the near future, hazardous waste exports are likely to increase further, despite intensifying efforts in recycling and source reduction. Recent articles from West Germany suggest that hazardous waste exports from that country will continue to occur and possibly even increase, regardless of the commitment of the government to manage its wastes within its borders (Frankfurter Rundschau 1990a; Frankfurter Rundschau 1990b; Frankfurter Rundschau 1990c; Die Zeit 1990; Frankfurter Allgemeine Zeitung 1990). To assess the potential future increase in hazardous waste exports is very difficult; however, existing proposals for waste scheme contracts might give an indication of the future scale. One such recent contract proposal made to Namibia, although not accepted, is illustrative: Under the terms of this contract, up to 5 million tons of hazardous wastes were to be imported each year for a period of 50 years. The company proposing this project would have paid US$ 1,000,000,000, besides several other payments, upon signing the license contract with the government. After this proposal was finally rejected, the offer was made to South Africa.

The increase in the export of hazardous wastes has several causes. First, given the past trend, the generation of hazardous wastes will at best remain constant or even further increase in industrialized countries, and it will certainly increase in newly industrialized countries. Some of these countries currently do not have the capacity to manage their own waste within their borders and, given the long lead times for hazardous waste facility siting, they will not have that capacity in the near future (see "Limited Disposal Facilities and Processing Capacity" in Chapter 3). In the EEC, facilities for safe disposal of hazardous wastes provide for a capacity of only 10 million tons/year, compared to the current generation of more than 30 million tons/year. The capacity shortage in the EEC, as a whole, is therefore more than 60 percent (EEB 1987, 126). As the capacity to manage hazardous wastes differs from country to country, in some countries the management shortage might be even higher.

Second, a number of newly industrialized countries, such as India, Malaysia, Korea, Mexico, Thailand, Indonesia, Philippines, and others, generate increasing amounts of hazardous wastes. The majority of these countries do not yet have an adequate waste management system, and other countries are very limited in their capability. If these countries want to ensure sound management of their wastes, they would have to export them to countries where proper management is possible. This, although very unlikely, would be necessary until they have built a sufficient waste management infrastructure.

Third, the efforts on the national and international levels, to coordinate a universal system of hazardous waste classification and definition, could easily result in higher volumes of hazardous wastes. As environmental standards become increasingly tighter—a trend observed in several industrialized countries (see "Higher Environmental Awareness and Tighter Regulations" in Chapter 3)—more waste will fall into the classification of "hazardous." Efforts to harmonize environmental standards are progressing, particularly within the EEC, which is working on the harmonization of environmental legislation within its member countries, through the Single European Act.[8]

Furthermore, the commitment of industrialized countries to decrease and finally terminate sea incineration of hazardous wastes will require additional management capacity, even though waste minimization and other disposal alternatives will compromise part of the decline of wastes assigned for sea incineration. In fact, 65 countries agreed at the 1988 London Dumping Convention, to ban all incineration of toxic wastes at sea by 1994. Since the volumes of wastes currently incinerated at sea are comparatively small, the need for new capacity will be modest. Figure 2-2 shows sea incineration for West Germany, which operates two incinerator ships in the North Sea.

More important, however, are the quantities of hazardous wastes that are dumped at sea. The volumes were close to 9 million tons in 1981 (World Bank 1989, 769). In some industrialized countries, the amount of hazardous wastes being dumped at sea has decreased over the past years. Figure 2-3 shows the decrease of sea dumping from 1980 to 1986 for the Federal Republic of Germany, which was the fourth largest dumper. The protection of the North Sea, the waste dump of many European countries and "one of the world's most polluted salt water bodies" (De Ligny 1990), has been on the agenda of many international environ-

---

[8]The Single European Act (SEA) was signed by the member states of the European Community in 1986. Its main provision is the creation of the Internal Market by the end of 1992. Article 100a (SEA) calls for the "harmonization of national provisions" for the creation of the market, which also affects environmental legislation. Article 130r par. 2 establishes the principles for environmental action. The Council of the EEC is acting on the proposal by the Commission, which stipulates that the Commission "must take as a base a high level of protection" (Koppen 1988).

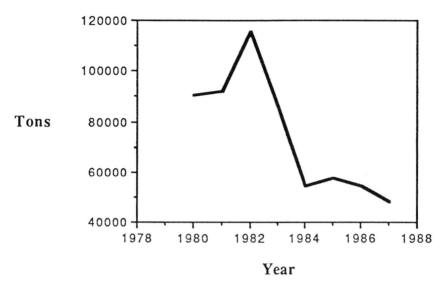

FIGURE 2-2.   FRG Sea Incineration—Volumes/Year.   Source: Unweltbundesamt. 1987. *Umweltjahresbericht.* Berlin: Umweltbundesamt.

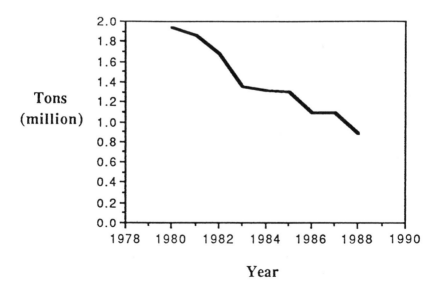

FIGURE 2-3.   FRG Sea Dumping—Volumes/Year.   Source: Unweltbundesamt. 1987. *Umweltjahresbericht.* Berlin: Umweltbundesamt.

mental negotiations. In 1987, the North Sea countries pledged to halt all industrial waste dumping by January 1, 1990, and, apart from the United Kingdom, all North Sea countries have met that deadline. The British government was criticized after its announcement that it would not meet this deadline and that it planned to phase out industrial waste dumping by late 1992 or early 1993. In the meantime, the United Kingdom continues to dump 205,000 to 270,000 tons of industrial waste annually (De Ligny 1990). During a follow-up conference in early 1990, nine Western European nations agreed on drastically curtailing emissions of four hazardous substances into the North Sea. Emissions of mercury, lead, cadmium, and dioxines will be reduced by 70 percent, effective in 1995. For 37 other toxic materials, the reduction target discussed was 50 percent, and on the issue of PCBs, it was agreed that they would be phased out by 1999 (De Ligny 1990). Currently, 45 tons of mercury, 3,500 tons of lead, and 95 tons of cadmium are dumped in the North Sea each year (De Ligny 1990).

The decrease of industrial waste dumping into the North Sea by almost all European countries, amounting to several hundred thousand tons each year, has created additional pressure on the waste management systems of these countries. As a result, many European countries face increased difficulties in their capability to manage their hazardous wastes. With reference to the continued British waste dumping, Mr. Patten, British Environment Secretary, said at the Conference on the Protection of the North Sea that "land-based means of disposal have not yet been established" (De Ligny 1990). This, in fact, is one of the reasons why hazardous wastes exports are defended as a necessary approach, to manage the increasing volumes of hazardous wastes.

The policies adopted to recycle hazardous wastes, to decrease and finally end sea incineration, and also to decrease the dumping of hazardous wastes at sea have several consequences. The pressure to dispose of the wastes abroad will increase, despite the fact that a growing share of hazardous wastes will be reduced at the source, recycled or reused.

This chapter described the context in which hazardous wastes are exported, as well as some of the problems of current legislation, monitoring, and enforcement. As the hazardous waste management industry expands rapidly, both in scale and international scope, it is also becoming more complex. The profits earned appear very tempting and attract new participants, who may not have the special expertise needed in this particular industry. The current practice of hazardous waste exports results in various other environmental and health-related consequences, which are discussed in Chapter 4. However, given these few characteristics, it can be argued that individual national approaches are insufficient to tackle the complex problems resulting from transboundary waste movements. A global policy approach is necessary, in order to effectively control waste streams, ensure sound waste management, and protect the human environment.

References

Benjamin, D. 1989. Returned to sender. *TIME*. August 28, 1989.

Centre Europe-Tiers Monde. 1989. *Nos Dechets Toxiques, L'Afrique A Faim, V'La Nos Poubelles*. Lausanne: CETIM.

Commission of the EC. 1989. *A Community Strategy for Waste Management*. SEC (89) 934 (final). Brussels: EC.

Conseil Europeen des Federations de L'Industrie Chimique (CEFIC). 1989. *CEFIC Position Paper on the Proposed Modification of the EEC Waste Directives* 75/442/EEC and 78/19/EEC.

De Ligny, R. 1990. Nine Western European nations agree on curtailing emissions of four hazardous substances into North Sea. Associated Press. *BBC summary of world broadcast*. March 7, 1990.

Deutscher Bundestag. 1988. *Antwort der Bundesregierung auf die "Kleine Anfrage der Fraktion der GRÜNEN," Drucksache 11/2644. Bonn*.

Die Zeit. August 3, 1990. Der wundersame Müllschwund. *Die Zeit* 32.

Erzeren, Ö. July 18, 1988. Verflucht sei die Giftmüllmafia. *Die Tageszeitung*.

European Environmental Bureau 1989. Final resolution of the international seminar "managing hazardous wastes: the unmet challenge." In *Toxic Terror: Dumping of Hazardous Wastes in the Third World*, pp. 126-28, Third World Network.

Exchange. 1988. Beyond our borders. *Exchange* 6(3):1-7.

Frankfurter Allgemeine Zeitung. July 20, 1990. Die Sondermüllentsorgung steht kurz vor dem Infarkt. *Frankfurter Allgemeine Zeitung*. July 20, 1990.

Frankfurter Rundschau. August 17, 1990a. Bayern erlaubt Müllexporte. *Frankfurter Rundschau*.

Frankfurter Rundschau. October 10, 1990b. Westliche Abfallhändler nutzen ganz Polen als riesige Müllkippe. *Frankfurter Rundschau*.

Frankfurter Rundschau 1990c. Der offizielle Abfall wird weniger—weil grobe Mengen auf illegalen Wegen verschwinden. *Frankfurter Rundschau*. July 27, 1990.

Fujiwara, M. and M. Tanaka. 1989. Municipal solid waste management in Japan. In *Konzepte in der Abfallwirtschaft 2*, ed. W. Schenkel, K.J. Thome-Kozmiensky. Bielefeld.

George A. August. 1988. False scent on the trail. *South*. pp. 38-39.

Gotoh, S. and K. Okazawa. 1989. Current status and future direction of hazardous waste disposal in japan. In *Konzepte in der Abfallwirtschaft 2*, ed. W. Schenkel, K.J. Thome-Kozmiensky. Bielefeld.

Greenhouse, S. March 23, 1989. UN conference supports curbs on exporting of hazardous waste. *New York Times*.

Greenpeace International. 1989a. *Waste Trade Negotiating Points*. Open letter to the delegates of the Organization of African Unity (OAU). Amsterdam: Greenpeace.

Greenpeace International. 1989b. International Trade in toxic wastes: policy and data analysis. Paper presented at International Export of U.S. Waste. Hearing before a Subcommittee of the Government Operations, House of Representatives, 100th Congress, July 14, 1988. Washington, D.C.: Committee on Government Operations.

Greenpeace International. 1989c. *Internationaler Mullhandel: Europa und der Mittlere Osten*. Washington, D.C.: Greenpeace.

Grefe, C. 1988. Blühendes Geschäft mit Giftmüllexport. *Euroforum* 11:7–8.

Hannequart, J.P. 1985. *Identification of Responsibilities in Hazardous Waste Management.* Paris.

Hansen, D. March 20, 1989. Hazardous Wastes: Curbs on shipment overseas urged. *Chemical and Engineering News.* p.6.

International Environment Reporter. 1988a, Concern about contracts. *International Environment Reporter.* 375 (BNA).

International Environment Reporter. 1988b. Italy recalls '*Karin B*', introduces remedial steps: UK, France spurn vessel. *International Environment Reporter,* 375 (BNA).

International Environment Reporter. 1989. International trade in hazardous wastes increases in 1988, university professors say. *International Environment Reporter.* February 1989.

Koppen, I.J. 1988. *The European Community's Environment Policy.* European University Institute Working Paper No. 88/328. Florence: European University Institute.

Martin, J.C. 1989. Testimony by Inspector-General of U.S. EPA. International Export of U.S. Waste. Hearing before a Subcommittee of the Government Operations, House of Representatives, 100th Congress, July 14, 1988. Washington, D.C.: Committee on Government Operations, pp. 12–27.

Mettelet, C. 1989. Production, traitement, recyclage et transferts transfrontaliers de dechets dangereux dans la Communite Europeenne. Agence Nationale pour la Recuperation et L'Elimination des Dechets—Les Transformeurs. Paper read at the workshop on Hazardous Waste Management Beyond 1992, 25./26. April 1989. Scientific & Technological Options Assessment (STOA), European Parliament, Brussels.

Ministry of Health and Welfare, Tokyo. 1990. *Solid Waste Management in Japan.* Tokyo: Ministry of Health and Welfare.

National Enforcement Investigation Center. 1988. *Enforcement Strategy Hazardous Exports.* Denver: U.S. EPA—Office of Enforcement.

Ogunseitan, S. 1988. Nigeria: The drums are gone but the poison remains. *Panoscope* (9):13–17.

Öko-Institut. 1988. Fern in der Türkei, wo der Giftmüll brennt....*Öko-Bericht: Sondermüll.* March 1988.

Organization of Economic Cooperation and Development. 1985. *Transfrontier Movements of Hazardous Wastes.* Paris: OECD.

OECD. 1988. *Decision of the Council on Transfrontier Movements of Hazardous Wastes.* 685th Session. C(88)90(Final). Paris: OECD.

OECD. 1989. *OECD Environmental Data Compendium 1989.* Paris: OECD.

Porterfield, A. and D. Weir. October 3, 1987. The Export of U.S. Toxic Wastes. *The Nation,* pp. 341–44.

Roelants du Vivier, F. 1988. *Les Vaisseaux du Poison: La route des dechets toxiques.* Paris: Sang De La Terre.

Scientific American. 1989. Dirty business. *Scientific American.* Special Issue. September 1989.

Schissel, H. 1988. The deadly trade: toxic waste dumping in Africa. *Africa Report,* pp. 47–49.

Sheppard, M. 1988. U.S. companies looking abroad for waste disposal. *Journal of Commerce*. July 20, 1988.

Synar. 1989. Opening statement of the Chairman of the Environment, Energy, and Natural Resources Subcommittee. Hearing before a Subcommittee of the Government Operations, House of Representatives, 100th Congress, July 14, 1988. Washington, D.C.: Committee on Government Operations, pp. 1-7.

The Amicus Journal. 1989. Toxic boomerang. *The Amicus Journal* 11(1):9-11.

Third World Network. 1989. *Toxic Terror, Dumping of Hazardous Wastes in the Third World*. Penang: Third World Network.

Time. August 28, 1989. *Time*.

Ueta, K. 1990. *The War on Waste in Japan*. Kyoto University Yoshida Honmachi, Kyoto, Japan.

Umweltbundesamt. 1987. *Umweltjahresbericht*. Berlin: Umweltbundesamt.

United Nations. 1987. *Environment Statistics in Europe and North America*. New York: United Nations.

United Nations Environment Programme. 1989. *Environmental Data Report*. Oxford: Alden Press.

U.S. Congress. 1989. International Export of U.S. Waste. Hearing before a Subcommittee of the Government Operations, House of Representatives, 100th Congress, July 14, 1988. Washington, D.C.: Committee on Government Operations.

U.S. News & World Report. November 21, 1988. Dirty jobs, sweet profits. *U.S. News & World Report*.

Vallette, J. 1989. *The International Trade in Wastes: A Greenpeace Inventory*. Washington, D.C.: Greenpeace International.

Vir, A.K. 1989. Toxic trade with Africa. *Environment, Science and Technology* 23(1):23-25.

World Bank. 1989. Hazardous Waste Management in Developing Countries. Technical Working Paper. Vol. III. Washington, D.C.: World Bank.

Yakowitz, H. 1985. Harmonization of specific descriptors of special wastes subject to national controls for eleven OECD countries. *Transfrontier Movements of Hazardous Wastes*. Paris: OECD.

Yakowitz, H. 1988. Identifying, classifying and describing hazardous wastes. *Industry and Environment* 11(1):3-10.

Yakowitz, H. 1989. *Possibilities and Constraints in Harmonizing National Definitions of Hazardous Wastes*. W/0837M. Paris: OCED.

3

Factors Affecting the Export of Hazardous Wastes

In the previous chapter, I reviewed the historical development of hazardous waste exports, the scope of these exports, and their potential future scale, as well as some of the particular features of current practices. In order to further understand the change in policy from hazardous waste disposal in the generating country to an increased reliance on exporting them, we need to examine the reasons behind this change. A number of factors contribute to the appeal of exporting hazardous wastes, regardless of the inherent risks. This chapter will examine these factors in more detail.

INCREASED OVERALL PRODUCTION OF HAZARDOUS WASTES

As indicated in the previous chapter, the volumes of hazardous waste exports have sharply risen in the past decade, and estimates for the future indicate a further increase. One of the underlying reasons for exports of hazardous wastes is the dramatic increase in their total production. The estimates of the extent of the global increase vary considerably: for the United States and the EEC, the largest generators, the OECD evaluated their increase in 1985 at 2 to 4 percent per year (Yakowitz 1985, 52). This figure contrasts sharply with certain U.S. government figures, which hover around 23 percent per year, for the increase of hazardous wastes production in that country, from 9 million tons in 1970 to 268 million tons in 1986 (U.S. General Accounting Office 1987). These latter estimates are shown in Figure 3-1. The line shows the production of hazardous wastes next to an exponential line indicating that the increase was exponential almost to the middle of the 1980s.

General economic growth is most likely the main reason for the increase in hazardous waste generation, and the developments in industrialized countries

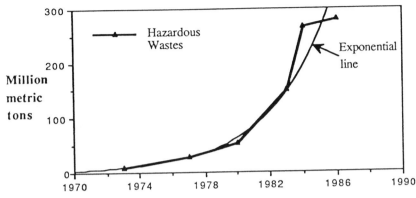

FIGURE 3-1.   Increase of Hazardous Waste Generation in the United States.   Different Sources: U.S. General Accounting Office 1987, EPA data and Estimates.

support this trend (Worldwatch Institute 1989, 70). The overall rise also stems from various other reasons, such as more stringent definitions and regulations covering wastes, which contribute to substantial increases in the volumes of waste having to be managed. In the past, newly industrialized countries produced only a minor share of world hazardous wastes, but will most likely contribute an increased portion in the future.

There would appear to be a correlation between the volume of hazardous waste generation and the tendency to export. In the United States, this is certainly true. The number of export notifications filed with the USEPA increased from 12 in 1980 to 465 in 1987 and 570 in 1988, as shown in Figure 3-2 (Synar 1989,1).

The intensified efforts to clean up old dumping sites represent additional sources of hazardous wastes to be managed. In the EEC, Denmark, The Netherlands, West Germany, and, increasingly, the United Kingdom and France are experiencing considerable efforts for soil clean up (European Environmental Bureau 1988, 5). These clean up activities will certainly continue over the next decades. Table 3-1 shows selected data on the number of contaminated sites. Estimates of the quantity of hazardous wastes that were improperly disposed of reach up to 90 percent in the United States (Findley and Farber 1988, 163). Given the present shortage of hazardous waste management facilities, these wastes will increase the pressure for exports. In fact, excavated, contaminated soil has been exported in the past. As an example, in 1986, 150,000 tons of contaminated soil from Belgium were dumped in the Netherlands (EEB 1988, 9).

There are also other interesting trends: the amounts of hazardous waste generated by large multinational corporations are not increasing, and in some cases are actually decreasing. Some companies are modifying their production processes to reduce waste and recycle or reuse their waste materials internally. Despite these

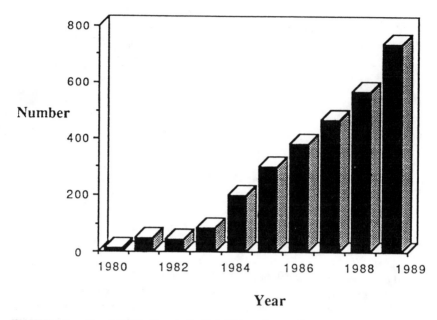

FIGURE 3-2.   Export Notifications to the U.S. EPA.   Source: U.S. Congress. 1989. *International Export of .S. Waste. Hearing Before a Subcommittee on Government Operations,* House of Representatives, 100th Congress, July 14, 1988. Washington, D.C.: Committee on Government Operations.

efforts, the total volume of hazardous wastes generated in industrialized countries has not declined. In West Germany, for example, the volume of wastes could not be reduced, since source reduction was overcompensated by increasing waste generation. Mr. Töpfer, the Minister for the Environment, Nature Protection and Nuclear Safety, currently talks about a "state of emergency" in the waste sector. In fact, despite the government's announcement that West Germany would not export hazardous wastes, 5 out of 11 states depend on waste exports (Schenkel 1990).

The steady increase in hazardous waste generation over the past decade is therefore clearly a primary reason to export wastes, especially in light of the limited disposal facilities in industrialized countries. The shortage of hazardous waste treatment and disposal capacity, another major contributor to an export oriented solution, is the topic of the next section.

LIMITED DISPOSAL FACILITIES
AND PROCESSING CAPACITY

Another reason why increasing generation of hazardous wastes has become a problem is the limited capacity for treating such wastes in the volumes produced.

TABLE 3-1.   Contaminated Sites

Country	Number of Contaminated Sites
The Netherlands	6,000
Denmark	3,000
West Germany	35,000
France	450
Belgium Flanders	400
Belgium Wallonia	8,000
United States	30,000–50,000*

Sources: Different sources: European Environmental Bureau (EEB). April 1988. *Soil Contamination through Industrial Toxic Dumps,* * R. Findley and D. Farber. Environmental Law in a Nutshell. 1988. St. Paul Minnesota: West Publishing Company, p. 162.

Within the EEC, the situation is characterized by an overall shortage of disposal[1] facilities, as compared to the identified needs. In West Germany, the government stated that the capacity shortage of incinerators is 600,000 tons/year (Neue Ruhr-Zeitung 1988), as compared to the 700,000 tons/year of installed capacity (Umweltbundesamt 1989, 468). Landfilling, the most widely utilized disposal method, is equally problematic in West Germany, as the number of operated landfills decreased from 1747 in 1980 to 1326 in 1984, or 24 percent in four years. Over the same period, the number of incinerators also decreased (Umweltbundesamt 1989, 456). The government has concluded that there are not enough hazardous waste disposal facilities and that the "situation might become worse in the next years..." (BMUNR 1987, 8). The problems from the former East Germany are not even taken into account here.

Other industrialized countries, such as Greece, Luxembourg, and Denmark, are small and "cannot afford" (Handley 1989, 10171-72) to build special disposal facilities, or their volume of hazardous wastes to be generated is so small that special facilities are economically inefficient. Furthermore, a country like The Netherlands bans landfills altogether because of its geological and hydrological conditions and a particularly high water table (Handley 1989).

In the United States, the situation appears even worse. Between 1982 and 1987, about 2700 landfills closed across the country (Helfenstein 1988, 788). The U.S. EPA estimates that "within a decade...more than half the (U.S.) states will have completely exhausted their landfill capacity and be unable to accept hazardous wastes" (Porterfield and Weir 1987, 456). Furthermore, the prospects for future

---

[1]In general, hazardous waste disposal is grouped in three categories: landfilling, physical-chemical treatment, and incineration. Depending on the substances, and so forth, it needs to be determined which form of disposal is the most appropriate. The analysis of the shortage of disposal facilities needs to reflect this distinction.

facilities being sited are not optimistic as "few if any new facilities are being built" (Helfenstein 1988, 775).

In an interview, a member of the Italian embassy in Caracas summarized the dilemma in which many industrialized countries have found themselves: "Italy has different hazardous waste disposal facilities, but not enough to dispose all the hazardous wastes we generate. This is the reason why we export" (Roelants Du Vivier 1988, 48). In sum, as shown for a few countries, industrialized countries currently lack the necessary capacity to manage the generated waste. This lack of management capacity is a contributing cause of hazardous waste exports.

HIGHER ENVIRONMENTAL AWARENESS AND TIGHTER REGULATIONS

The public awareness of the environment has changed considerably in the past three decades. Two decades ago, pollution was the main concern; at present, people are alert to many other environmental issues, such as the changing atmosphere and climate, threats to biodiversity, depletion of natural resources, and the relationship between the environment and development.

Between 1981 and 1984, the OECD conducted a survey on environmental issues that showed a significant common pattern of environmental awareness in industrialized countries, namely the United States, Japan, and European countries. The survey indicated that 43 percent of the public was concerned about industrial waste disposal, and 38 percent about transfrontier pollution. It also indicated that the public opinion supporting environmental improvements has remained remarkably strong over time (UNEP 1988, 173-74). Environmental disasters in many industrial countries from *Love Canal* to *Seveso* have alerted and frightened the public. The story of *Love Canal* has been told many times and does not to be repeated here (Levine 1982). For many people, however, it was the first time that they had heard the term *hazardous wastes*. The role of the mass media, which reported on and documented these accidents, has reinforced that concern and has been instrumental in promoting environmental awareness.

As a result of this increasing public awareness, pressure has grown to clean up the environment. At the same time, more and more environmental organizations have been formed to lobby for legislative and regulatory action. In the United States, for example, the Council on Environmental Quality (CEQ) was founded in 1969 and the EPA in 1970. These early environmental activities led to the National Environmental Policy Act and other legislation concerning air and water pollution, as well as hazardous wastes.[2] The developments in other

---

[2] An overview of national waste legislation and control of the transboundary movements of hazardous wastes is given in Chapter 5.

industrialized countries, especially in Europe, have been similar to those in the United States.

While hazardous waste has maintained a "high public profile," (Burns 1984, 188) public perceptions of hazardous waste management show that little credibility is given to government assurances about the effectiveness of state regulations. Industry, for example, is believed to be reluctant to accept new regulations requiring environmentally-sound waste management practices, and current waste handling methods are seen as motivated only by profit considerations (Burns 1984, 188).

As a result of these perceptions, as well as the higher awareness, there has been a general increase in public participation in the decision-making process concerning environmental issues, and, in particular, regarding hazardous waste management facility (HWMF) siting. Despite the participation in the process, public opposition to the siting of these facilities has become very strong. This opposition stems from the fears of threats to health and life, and from numerous other social, psychological, and economic factors associated with hazardous wastes (Pedersen 1989). Presently, it is very difficult in several industrialized countries to site new landfills or incinerators, and the situation has been described as an "environmental emergency." In fact, this is one of the reasons why hazardous wastes are being exported.

Moreover, as the opposition to hazardous waste management facility siting has grown, the leadtime for new facilities has reached several years, adding to the existing pressure on waste management. In the late 1970s and early 1980s, a second wave of public concern over environmental issues has led to more stringent laws and higher standards for hazardous waste management.[3] For instance, the Resource Conservation and Recovery Act (RCRA) banned numerous hazardous wastes from being disposed of in landfills, and, according to the EPA, has led to an increase in exports (Hansen 1989, 6). In the EEC, where hazardous waste laws among member states are still very diverse, some member states have tightened their laws, while others have not, which has provided additional incentives for the former to export hazardous wastes (Roelants du Vivier 1988, 77). West Germany is again a good example. Recent legal changes increased the number of substances defined as hazardous from 80 to approximately 350. The law also requires a specific management technique for each waste, forcing waste generators and disposers to follow more precise instruc-

---

[3]In the United States, the Resource Conservation and Recovery Act (RCRA) was passed in 1976, and amended in 1984. The "Comprehensive Environmental Responsibility, Compensation, and Liability Act (CERCLA) was passed in 1980 and amended in 1986. In other industrial countries in western Europe, which generate large amounts of hazardous wastes, the developments in amending environmental laws were similar.

tions as to what management option they choose. At the same time, the new law also strengthens pollution standards for landfills and incinerators (Harenberg-Verlag 1990). With these regulatory improvements, the waste management industry faces a two-fold problem: the pressure to manage more hazardous waste, while having less disposal capacity as a result of more stringent disposal standards.

Higher regulatory standards and more stringent laws coupled with high public awareness have affected industry practices in dealing with hazardous wastes. Facing public pressure and less disposal capacity, industry has increasingly moved to export wastes abroad. However, the exporting of wastes should not be seen as an insignificant operation. In recent years, hazardous waste management—and particularly recycling—has increasingly become an international business. This can be seen through the investments of large waste management companies abroad and will be even more so with the creation of the Internal Market of the European Community. Firms like Waste Management or BFI, from the United States, have invested in France, Italy, and Holland and probably already control 10 percent of the waste market in these countries (Schenkel 1990). European companies, like BASF, DuPont, ICI, Bayer, and others, similarly engage in international waste management activities. These activities may also have positive effects on the management practices as more competition, standardized practices, and the application of state of the art technology become part of the business.

HIGHER DISPOSAL COSTS

In the previous sections, I discussed how the volumes of hazardous waste increased, while the disposal capacity for hazardous waste management decreased over the same period, and the regulations concerning such wastes were strengthened and tightened. All three of these transitions have contributed to an overall increase of disposal costs.

The financial costs of hazardous waste disposal have been affected in several specific ways:

1. Certain wastes that under previous regulation, were deemed nonhazardous, are, under new regulation, classified as 'hazardous' and thus require special management.
2. Certain hazardous wastes that could have been disposed of in landfills before cannot be landfilled any longer, but must be incinerated or treated by a physical-chemical process.
3. Tightened emission standards and regulations for landfills have resulted in the closure of disposal facilities.

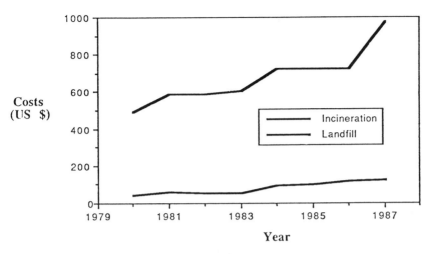

FIGURE 3-3.   Disposal Costs in the United States Per Metric Ton.   Source: U.S. Environmental Protection Agency, Office of Policy Analysis. 1984. Survey of Selected Firms in the Commericial Hazardous Waste Management Industry. Update, 1985, and 1986/87. Washington, D.C.: EPA.

4. The production of hazardous wastes has increased, against a background of limited disposal capacity.
5. Insurance and liability requirements have risen. The result of all these changes has been higher costs pertaining to waste management.

Figure 3-3 shows the increase in hazardous waste disposal costs for landfilling and incineration in the United States. According to the EPA, the prices for landfills have increased "sixteenfold" since the early 1970s and currently vary from US\$ 250 to US\$ 350 per ton of hazardous waste (Worldwatch Institute 1988, 3).[4] For incineration, the increase of average costs is from US\$ 500 in 1980 to US\$ 1,500 per ton in 1989 according to UNEP (Cirulli 1989, 10). The prices in Europe correspond to those in the United States (Mettelet 1989, 7). From these figures, it can be concluded that hazardous waste generators have a strong financial incentive to look for cheaper disposal methods. If the total cost of packaging, transport, and management abroad is cheaper than that of legal disposal in the country of generation, there is an incentive to export the wastes (MacNeill 1985, 8). A very recent study suggests that the average cost differential between landfilling and incinerating hazardous wastes, including organic, heavy

---

[4]The prices are given for *metric tons*, although many sources do not specify metric or U.S. tons.

metals, or cyanidic components, is likely to be more than 200 ECU (Yakowitz 1989, 4).[5]

Assuming that the costs for administration, packaging, labelling, transport, and insurance are US$ 100 per ton of exported waste, the marginal cost savings for generators is at least US$ 100 (Yakowitz 1989, 4). The volumes of hazardous waste crossing frontiers worldwide, as estimated in the previous chapter, are from 3 to 3.5 million tons per year. Based on these data, the marginal savings amounts to approximately US$ 300 to US$ 350 million per year for the waste trade among industrialized countries alone.

A similar analysis undertaken by the OECD gives annual marginal savings between 200 and 300 million ECU for OECD Europe only. The study further concludes that the "trade" in hazardous wastes reaches 500 million ECU per year (Yakowitz 1989, 4).

Disposal costs in less and least developed countries (LLDC), or newly industrialized countries (NIC), whose disposal method is almost exclusively landfilling, are however *lower* than within OECD countries. Analyzing past export schemes to developing nations show that prices paid were from as low as US$ 2.50 per ton up to about US$ 40 (Cirulli 1989, 10). If the final disposal in LLDCs were even US$ 50 per ton, an additional saving of US$ 200 to US$ 300 could be gained. The result would be an additional marginal saving of US$ 600 to US$ 700 million, based on the same volume of waste. The total potential financial savings from the worldwide export of hazardous wastes to LLDCs and NICs, as compared to incineration in the country of generation, is thus in the range of US$ 900 million to US$ 1.05 billion per year. The export of hazardous wastes would therefore appear to be a very profitable business.

Associated with the higher economic costs of waste management are risk mitigation and/or risk avoidance. The export of hazardous wastes does not clearly result in an overall mitigation of risks; rather, it represents a transfer of these risks to other countries. Two aspects are particularly relevant to any discussion of costs and risk: liability and insurance. These issues are discussed in detail in the policy

---

[5]The following calculation shows the rationales behind different disposal options. It is based on the following considerations: (1) The kinds of hazardous waste that are considered for export and that are actually exported are highly hazardous and, in general, require special disposal, such as incineration. The costs of this disposal method are very high and export thus gives the largest marginal savings for generators. (2) The calculation does not reflect savings for disposing of PCBs, whose incineration costs are up to US$ 1,500 per ton (Cirulli 1989, 10). (3) The calculation does not include financial savings resulting from sea incineration and sea dumping, a volume of more than 10 million tons/year, which would incur costs in the magnitude of billions of US$ per year if disposed of properly. Finally, (4) the marginal savings are based on transfrontier waste movements occuring *within* the OECD, in which the export of hazardous waste can be considered a "trade," and, as calculated, represent the *difference* between landfilling and incineration.

analysis in Part 2; however, in general it can be said that the laws regulating hazardous wastes in many industrial countries now have provisions attributing liability to the generator and the disposer of hazardous wastes. As has been clearly shown in previous waste accidents and spills, the financial burden for liability can be very large, depending both on the kind of accident and on the specific liability provisions. Insurance is thus a necessity, but could be extremely expensive or even difficult to obtain (MacNeill 1985, 11). Currently, the export of hazardous wastes represents an option to avoid liability and insurance costs and is therefore another incentive for waste generators to export.

INTERNATIONAL DEBT AND ECONOMIC PROBLEMS

For developing or newly industrialized countries, the practice of importing waste appeared to represent an additional source of income in the past and was therefore considered "very profitable" (Yakowitz 1989, 4). Particularly in light of the tremendous debt burden many LLDCs and NICs face, the import of wastes seemed to constitute a tempting opportunity.

The debt many LLDCs and NICs are facing today has been built up over a considerable period of time, starting from the early 1970s. Easy credit, market fluctuations, the general fall in real prices for the primary commodities on which many of these countries depend for revenue, and widespread weakness in managing their economies are all contributing factors to the debt calamity. It is safe to say that the debt burden is now one of the most serious, and indeed crippling, problems for many LLDCs and NICs (see Figure 3-4). One unexpected consequence has been that the transfer of financial resources from industrialized countries to LLDCs and NICs was turned around. According to the World Bank, since 1984 the net transfers from industrialized to developing countries have been negative. In other words, LLDCs and NICs pay more money to industrial countries than the amount they receive in international financial assistance (BUND 1988, 3-5). Figure 3-5 shows the transition period for the net transfer from the early to the mid-1980s. Some LLDCs and NICs were so desperate to obtain foreign exchange that they engaged in the hazardous waste export/import business. The case of the small West African country, Guinea Bissau, is illustrative of the real scope of the problems some of these countries face on the one hand, and the tempting offers made, on the other hand.

Guinea Bissau, with a population of 300,000 and a Gross National Product (GNP) of US$ 150 million, is one of the poorest countries in Africa. Its foreign debt is approximately US$ 300 million, and its debt-service ratio[6] is 1900 percent. This

---

[6]The debt-service ratio represents the service payments (amortization plus interest) on foreign debt as a percentage of the exports of goods and services.

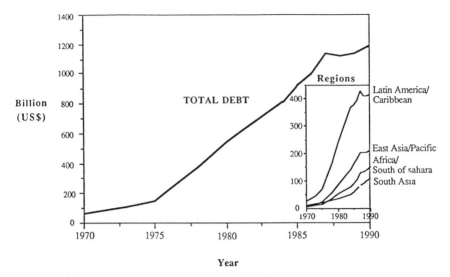

FIGURE 3-4.   Net Transfer to and from all Developing Countries.   Source: World Bank. 1989. World Debt Tables 1970-79, 1984-1991, Vol. 1-3. Washington D.C.: The World Banks.

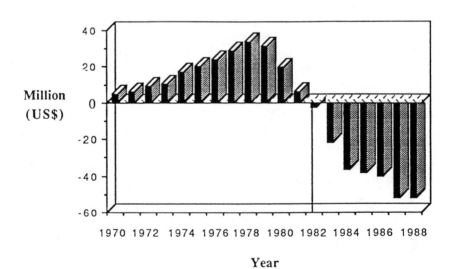

FIGURE 3-5.   Growing Debt of Developing Countries.   Source: World Bank. 1989. World Debt Tables 1970-79, 1984-1991, Vol. 1-3. Washington D.C.: The World Banks.

ratio is nearly 15 times greater than the 150 percent considered by the OECD at the margin to file bankruptcy (Michler 1988, 5). Guinea Bissau was offered a contract by a Swiss firm, "Intercontract," to import 50,000 tons of hazardous wastes per year over a period of 15 years. For the total volume of up to 15 million tons, the government was offered a sum of US\$ 600 million, twice the country's foreign debt and about 25 times greater than current export revenues (EPD 1988, 22). The Guinea Bissau contract was initially accepted, but later canceled due to public pressure.

Offers like this one were indeed so tantalizing that many LLDCs and NICs accepted them. As a reaction to public outcry and protest in the late 1980s, hazardous waste merchants were not relenting, but "raising the stakes high enough that cash-trapped African countries may find the overtures of the toxic waste merchants too tempting to ignore" (*African Business* 1989, 16). Such contracts offer not only substantial sums of cash, but also package toxic waste exports as infrastructure development schemes (Hall and Karliner 1988, 4). A typical contract of this sort incorporates the construction of a deep-water harbor, two 100,000 ton/year waste burning plants capable of generating electricity, construction of railway tracks, and so on (*African Business* 1989, 17). In addition, such projects bring badly needed employment opportunities and stimulate local and regional economic growth. In light of these advantages, hazardous waste export deals became a *booming business*.

The debates on hazardous waste exports in many governments of industrialized countries have raised the issue of economic dependency of LLDCs and NICs as a result of their foreign debt. There are two main positions: The first holds that industrialized countries are partially responsible for the economic crisis in LLDCs and NICs and thus bear some responsibility for finding a solution to the debt problem. This position also questions the acceptability of hazardous waste exports from an ethical standpoint, especially given the desperate economic conditions in these countries. The other position stresses the sovereignty[7] and independence of LLDCs and NICs and argues that whether or not LLDCs and NICs engage in hazardous waste imports should be solely their decision.

Carlos Pimenta, Portuguese deputy to the European Parliament and former Minister of the Environment in Portugal, strongly defended the former with the following plea:

> ...some will say that preventing sovereign nations from freely accepting toxic waste constitutes a new form of paternalism or colonialism: This is not true! On the contrary, it is shameful on the part of Europe or of the other industrialized nations from the

---

[7]The issue of national sovereignty has been raised in a number of debates concerning environmental protection, in particular, transfrontier pollution problems (Piddington 1989).

West and East, to have allowed that African, Pacific, and Caribbean countries have reached such a state of poverty and economic dependance, that they are forced to accept a meager amount of money in compensation for taking the wastes we send them (Roelants du Vivier 1988, 97–98).

Both positions have a valid argument; however, I think they should not be used against each other, as responsibility from both sides is needed to find a solution.

DISPOSAL AT THE MOST APPROPRIATE LOCATION

Another factor to consider in waste management is the potential economic value of waste. Certain precious metals and minerals can be recuperated from some wastes, and therefore represent an economic benefit. Recycling and reuse of hazardous wastes are on the increase in industrialized countries, where some managers strive to increase plant efficiency and minimize waste disposal costs.

Some countries' geo-ecological situation does not allow the disposal of hazardous waste in landfills. One example is the Netherlands, which in the past has chosen to export. In other cases, the best available technology for hazardous waste disposal might not be available or affordable. Developing and newly industrialized countries often do not have access to state of the art waste management technologies, which are patent protected. The license fees are often too high and thus unaffordable, due to scarce financial resources and/or other economic priorities. Many countries, among them several developing Eastern European nations, find themselves constrained to choose between investing in environmental protection or in general economic development. The trade-offs often result in infrastructure development investments. Countries generating only a moderate amount of hazardous wastes often consider export the best political and economic solution, and one that does not compromise on environmental protection. Their production of PCB-contaminated waste might, for instance, be too little to operate an incinerator economically. In such a case, they might invest in a joint venture with a neighboring country and build an incinerator together. Alternatively, they may decide to export their wastes.

To search for the optimal location for environmentally sound management of hazardous wastes is a complex adventure involving many interdependent economic, social, and political factors. In 1984, UNEP conducted a study on the disposal of hazardous wastes in LLDCs and NICs. The study points out that compared to countries lying in the more temperate zones of the earth, countries within or near the tropical zone are the worst situated for hazardous waste disposal. Substances that are disposed of in landfills are subjected to intense tropical rains, and it is quite common that the landfills overflow due to torrential rains (Roelants du Vivier 1988, 83).

The proposal to dispose of hazardous wastes at the most appropriate location is a logical approach. Nevertheless, the search for such a location is hampered by numerous factors. Not only are the wastes very different in their chemical-physical composition, but also different environmental, economic, and sociopolitical policies in developing and industrialized countries render the task of finding an optimal universal location virtually impossible.

CONCLUSION

This chapter discussed the most important factors affecting the export or import of hazardous wastes. The main reason for waste exports is that legal disposal has become increasingly costly in the industrialized countries. The generators of hazardous wastes naturally seek the least costly legal disposal of their wastes, making export a prime choice, if available. For LLDCs and NICs, the problem is quite different: the factors bringing hazardous waste imports on their agendas are primarily their foreign debt and their general economic crises. Much less important is the need to dispose of an increasing amount of self-generated hazardous wastes, or a waste-related technology transfer.

References

African Business. March 1989. Toxic waste merchants 'offer $ 2bn'. *African Business*, pp. 10-17.

Bundesministerium für Umwelt, Naturschutz und Reaktorsicherheit (BMUNR). 1987. *Umweltbrief: Abfallgesetz*. Nr. 36. Bonn: BMUNR.

Bund für Umwelt und Naturschutz. 1988. *Dritte Welt: Entwicklungen gegen die Umwelt?!*. Bundesjugendkongress des Bund für Umwelt und Naturschutz. Deutschland.

Burns, M.E. 1984. Striking a Reasonable Balance. In *Hazardous Waste Management, In whose Backyard?* ed. Harthill, M.

Cirulli, C. 1989. Toxic Boomerang. *The Amicus Journal* 11(1):9-11.

Der Standard. August 14, 1989. Agrarhandel: Die Dritte Welt hat wenig Chancen. *Der Standard*.

European Environmental Bureau (EEB). April 1988. *Soil Contamination through Industrial Toxic Dumps*.

European Parliament. 1987. Report on the waste disposal industry and old waste dumps. Rapp. Roelants du Vivier. Doc. A 2-31/87.

Evangelischer Pressedienst (EPD). July 14-15, 1988. *Entwicklungspolitik*.

Findley, R.W. and D.A. Farber. 1988. *Environmental Law in a Nutshell*. St. Paul: West Publishing Company.

Hall, B. and J. Summer Karliner. 1988. Garbage imperialism. *EPOCA Update*.

Handley, J. 1989. Hazardous waste exports: a leak in the system of international legal controls. *Environmental Law Reporter* 19(4):10171-182.

Hansen, D. March 20, 1989. Hazardous wastes: curbs on shipments overseas urged. *Chemical and Engineering News*, p. 6.

Harenberg Verlag. 1990. Giftmüll. In *Aktuell 91. Das Lexikon der Gegenwart*. Harenberg Lexikon-Verlag.

Helfenstein, A. 1988. U.S. controls on international disposal of hazardous wastes. *The International Lawyer* 22(3):775–90.

Levine, G. A. 1982. *Love Canal: Science, Politics, and People.*

MacNeill, J.W. 1985. Policy Issues Concerning Transfrontier Movements of Hazardous Waste. *In Transfrontier Movements of Hazardous Wastes*, pp. 7–12. Paris: OECD.

Mettelet, C. 1989. Production, traitement, recyclage et transferts transfrontaliers de dechets dangereux dans la Communite Europeenne. Agence Nationale pour la Recuperation et L'Elimination des Dechets—Les Transformeurs. Paper read at the workshop on Hazardous Waste Management Beyond 1992, April 25–26, 1989. Scientific & Technological Options Assessment (STOA), European Parliament, Brussels.

Michler, W. 1988. Afrika: Giftmüllkippe für die Reichen. *Dritte Welt Presse* 1(5):5.

Neue Ruhr-Zeitung August 4, 1988. Bonn: 'Giftmüll-Tourismus' wird strenger kontrolliert. *Neue Ruhr-Zeitung.*

Pedersen, J. 1989. *Public perception of risk associated with the siting of hazardous waste treatment facilities.* Dublin: European Foundation for the Improvement of Living and Working Conditions.

Piddington, K.W. 1989. Sovereignty and the environment. *Environment.* 31(7):18–20, 35–39.

Porterfield, A. and D. Weir, October 3, 1987. The export of U.S. toxic wastes. *The Nation*, pp. 341–44.

Roelants du Vivier, F. 1988. *Les Vaisseaux du Poison: La route des dechets toxiques.* Paris: Sang De La Terre.

Schenkel, W. 1990. Stand und perspektiven der abfallwirtschaft in der Bundesrepublik Deutschland. In *Abfallwirtschaft und Deponietechnik* '90, ed. K. Wiemer, pp. 31–52. Kassel: Universität Kassel.

Synar. 1989. Opening Statement of the Chairman of the Environment, Energy, and Natural Resources Subcommittee. *Hearing before a Subcommittee of the Government Operations*, House of Representatives, 100th Congress, July 14, 1988, pp. 1–7. Washington, D.C.: Committee on Government Operations.

Umweltbundesamt. *Daten zur Umwelt* 1988/89. Berlin: Umweltbundesamt, FRG.

United Nations Environment Programme. 1988. *The State of the Environment.* Nairobi: UNEP.

U.S. Congress. 1989. *Hearing before a Subcommittee of the Government Operations*, House of Representatives, 100th Congress, July 14, 1988, pp. 1–7. Washington, D.C.: Committee on Government Operations.

U.S. Environmental Protection Agency, Office of Policy Analysis. 1984. Survey of Selected Firms in the Commercial Hazardous Waste Management Industry. Update, 1985, and 1986/87. Washington, D.C.: EPA.

U.S. EPA. Office of Policy Analysis. 1986. *1985 Survey of Selected Firms in the Commercial Hazardous Waste Management Industry.* Washington, D.C.: EPA.

Worldwatch Institute. 1988. *State of the World 1988.* New York: Norton & Company.

Worldwatch Institute. 1989. *State of the World* 1989. New York: Norton & Company.

Yakowitz, H. 1985. Harmonization of specific descriptors of special wastes subject to national controls for eleven OECD countries. In *Transfrontier Movements of Hazardous Wastes*, pp. 50-81. Paris: OECD.

Yakowitz, H. 1989. *Monitoring and Control of Transfrontier Movements of Hazardous Wastes: An International Overview*. W/0587M. Paris: OECD.

4

Consequences of Hazardous Waste Exports

The export of hazardous wastes can have a serious adverse effect on the environment and ecology. The potential for ecosystem damage is substantial if hazardous wastes are disposed of in an uncontrolled manner. Many waste export schemes have been managed in uncontrolled ways and have thus posed hazards to human life and health. The health and environmental effects of unsafe disposal of hazardous wastes are caused by their toxicity, flammability, corrosivity, or reactivity.

Furthermore, if wastes are dumped in uncontrolled landfills in the importing countries, the dump sites will have to be cleaned up eventually, to comply with law suits or new legislation mandating clean up. The cleanup and final disposal costs, in addition to compensation for poisoned victims, may together present unexpectedly high financial burdens to importing countries.

Aside from the health, environmental, and financial consequences, past hazardous wastes export practices have also triggered more or less severe political and diplomatic problems, and contributed to the deterioration of relations between some industrialized countries and LLDCs and NICs.

A policy covering hazardous waste exports must also take into consideration other issues in waste management, such as how the reuse and recycling of waste generated in industrialized countries will be affected, and, even more importantly, what the consequences will be on efforts to promote waste reduction at the source. Such a comprehensive perspective is important because many industrialized countries have adopted policies to give source reduction, reuse and recycling a priority in their hazardous waste policies. Various policy options are discussed in Part 2. This present chapter will examine some of the potential risks.

DAMAGE TO THE ENVIRONMENT
AND ECOLOGY

Toxic substances can enter the environment and possibly cause damage through soil, water, or the atmosphere. Figure 4-1 shows these physical and biological routes of transport, their release from the disposal sites, and their potential for human exposure. The importance of each pathway depends not only on the physical, chemical, and biological properties of the waste, but also on the characteristics of the disposal site and the underlying geology (World Bank 1989, 24).

Uncontrolled disposal of hazardous waste is as likely to cause damage to the environment and ecology in LLDCs and NICs as it has in industrialized countries: the problems encountered in places such as Love Canal in the United States, Lekkerkerk in the Netherlands, and Seveso in Italy, to name only a few, can all too easily be replicated in Africa, Asia, and Latin America (WCED 1987, 211). Up to now, there is no information suggesting that waste disposal in LLDCs and NICs is environmentally safer than in industrialized countries. On the contrary, these countries have less capacity to ensure correct treatment of wastes, and one study

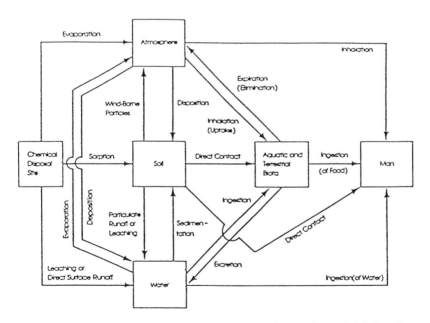

FIGURE 4-1.   Physical and Biological Routes of Transport of Contaminants, their Release from the Disposal Site, and Potential for Human Exposure.   Source: World Bank. 1989. The Safe Disposal of Hazardous Wastes: The Special Needs and Problems of Developing Countries, ed. Batstone, R. et al., World Bank Technical Working Paper No. 93. Vol. I. Washington, D.C.: World Bank.

suggests that, for climatic and other exogenous reasons, waste disposal in tropical zones is less appropriate than in industrialized countries (see Chapter 3).

Contamination of Soil

Much of the Earth's land surface is unsuitable for agriculture. Only 11 percent, or 14 million square kilometers, of the total land area is fertile enough for agriculture. It is this part of the Earth that is used for world food production. Furthermore, it takes up to 12,000 years to build up a layer of top soil sufficient for agriculture (Clark and Palmer 1983, 14). This soil can, however, be damaged or lost by water contaminated by leaking disposal sites or other forms of environmental distress, such as desertification. Currently, some 2 million hectares of land are lost per year to agriculture due to toxification alone (Clark and Palmer 1983, 14). Also, as pointed out in Chapter 3, in 6 out of the 24 OECD member countries alone, there are more than 80,000 contaminated waste sites waiting to be cleaned up. Many of these sites have the potential to contaminate and damage fertile soil. It therefore seems obvious that these dump sites should be cleaned up. More importantly, no new dumps should be created in either industrialized or developing and newly industrialized countries.

Pollution of Ground Water

Landfills may contaminate groundwater through leaching or direct overflow caused by too high a water content in the landfill. There have been several hundred cases of contamination of wells from hazardous wastes, many of them arising from improper waste management (UNEP 1983, 7). The Rhine and Mississippi rivers, for example, have been continually polluted by industrial wastes. Along the Rhine there are numerous communities whose wells are connected to the river. In the past, pollution accidents have forced the closure of these wells (UNEP 1983, 7).

In LLDCs and NICs, most industrial production is concentrated in congested areas. Access to waterways is required for either production processes or pollution discharge or both. Solid wastes generated at these plants are generally landfilled close to the production sites, and these landfills are in many instances surrounded by poor neighborhoods (Yakowitz 1989, 6). The concentration of industrial activities and human settlements in coastal regions and estuaries—seven out of every ten people on Earth live within 50 miles of a coast (Blackburn 1986, 151)—increases the risk of human exposure to hazardous wastes or to contaminated ground water, as compared to rural villages. In fact, wastes from past export schemes have often been disposed of by or near rivers and oceans because of the easy accessibility for the ships transporting the wastes. These ocean liners are bound for deep sea ports that are often the center of large settlements, such as Dakar, Bissau, Conakry, Freetown, and Lagos in West Africa. Not surprisingly, some of these cities have been the focus of proposed hazardous waste export schemes.

Contamination of Rivers, Sea Resources, and Coastal Regions

Contaminated surface waters may also poison groundwater. Wastes disposed close to rivers and oceans may contaminate these water bodies through runoff of polluted surface water, and can thus affect the entire water table. Heavy rains in the tropical rain belt, combined with high humidity, may increase the risks of leaching and chemical reactions. Finally, waste disposals near or beside oceans are more vulnerable to particular or total destruction by hurricanes and flooding.

Given the already endangered conditions of many rivers and coastal zones (WCED 1987), uncontrolled discharge of industrial or other wastes can seriously affect and further damage these water systems. LLDCs and NICs, which are in the process of formulating and establishing waste policies, could well learn from the experience of the industrialized countries. Uncontrolled waste discharge and other unchecked pollution have caused significant environmental damage, such as the eutrophication of the Great Lakes (United States), the flooding ashore of medical wastes in New England (United States) in the summer of 1989, the profusion of algae in the Riviera (Italy), and the poisoning of the Rhine river (Europe) by the Sandoz plant fire. The consequences were multiform, and included direct and indirect hazards to humans and animals, the marine life in general, and serious economic losses (such as from lost tourist and other revenues).

Air Pollution

Contaminants from disposal sites can become volatile and cause serious air pollution. The open waste dumps often found in LLDCs and NICs have vaporization rates greater than the covered landfills common in industrialized countries. As a result, the quantities of volatile wastes released into the atmosphere at landfill sites can be substantial on an annual basis (World Bank 1989, 26).

Depending on the substances and their ability to react, wind-blown dispersal of contaminants can carry toxic matter to inhabited areas. Certain wastes (for instance, asbestos) are particularly susceptible to wind-blown dispersal (World Bank 1989, 26). Some of the reports on waste exports to LLDCs and NICs have disclosed hazards from gases and other forms of air-transmitted pollution (Third World Network 1989).

Threats to Precious Resources and Biodiversity

Toxic substances contained in hazardous wastes not only affect life processes in the flora and fauna, but also damage and destroy natural resources—above all, water, one of the most vital resources for life on this planet. As pointed out in one study, "severely polluted water may still flow in a river but provide no fish to catch, and no water for drinking and irrigation" (Wesley 1971, 55–57). The supply of sufficient drinking water is probably one of the most challenging tasks facing mankind, as 1.7 billion people lack access to clean water (UNEP 1986, 109). Furthermore, water, air, and soil contamination may pose a serious threat to the diversity of vegetation and animal life.

Leaching from waste disposal sites has caused irreparable damage in some areas of industrialized countries. In LLDCs and NICs, such accidents could be even worse, because these countries are in a more vulnerable state of development. These countries often have little or no working regulatory or monitoring system for pollution, and adverse environmental effects are therefore difficult to track down and to counter. Furthermore, the LLDCs and NICs lack the technical and financial resources of the industrialized countries to undertake such operations.

THREATS TO HUMAN HEALTH AND LIFE

Improper hazardous waste disposal can harm human health and life. Toxic substances can enter the human body by inhalation, by ingestion, and through direct skin contact (UNEP 1986). A major source of exposure is the food chain, including drinking water (see Figure 4-1). The hazardous wastes can be pathogenic, mutagenic, cytogenic, and carcinogenic to humans, and/or affect fetal and neonatal growth and development (World Bank 1989, 33). The following four sections briefly discuss these potential hazards to the human environment.

Acute Reactions

Compared to several thousands of different chemical substances on the market, scientific information about the short-term effects of chemical hazards to human health is limited. The effects of arsenic exposure can be immediate and obvious when the dose is large, resulting in illness or even death. Acute reactions after exposure to toxic substances have reportedly been headaches, nausea, dizziness, and discomfort (UNEP 1986, 117). Other acute reactions include irritation of eyes and skin. In a toxic waste export scheme in the Dominican Republic:

> wastes containing antibiotics and fish oil were imported for use as cattle feed and fertilizer from a U.S.-based Abbott Laboratories' facility in Puerto Rico, where pharmaceutical dumps are full....According to Dr. Antonio Thomen, director of the Domini-

can Institute for Bioconservation, ingestion by humans can [*sic*] cause hormonal disorders, birth defects and severe intestinal illnesses, particularly among children (Porterfield and Weir 1987, 343).

The PCB-contaminated wastes exported to Nigeria and described in detail in Case 3 resulted in "dockworkers becoming paralyzed or suffering severe chemical burns, premature births and 19 deaths from contaminated rice" (*U.S. News & World Report* 1988, 55).

The scope of the danger posed by a chemical after it enters the environment is mainly a function of its toxicity and the extent of human exposure (Postel 1988, 120).[1] Acute reactions in human beings are often the first health-related signs that toxic substances may have entered the environment.

Carcinogens

Chemicals are considered carcinogenic if an exposed population shows "abnormally" higher incidences of cancer compared to an unexposed population. Many chemicals "cause cancer because they are mutagenic, producing mutations in cells other than those which are transmitted from generation to generation" (Consumers' Association of Penang 1989, 98). Once again, *Love Canal* is an example of a hazardous waste dump where tests have shown that "some residents had damaged chromosomes," raising the spectre of cancer (*Third World Network* 1989, 16).

This description of possible damage from exposure to toxic substances should not only be seen in the context of safety standards, enforcement of regulations, and advanced health monitoring found in industrialized countries. Special attention must be paid to the LLDCs, where adverse health effects may not be monitored or counteracted because of an insufficient infrastructure. The risks to the population of contamination from improper disposal of hazardous wastes are comparable to those prevalent in sophisticated technology transfers from industrialized countries to LLDCs and NICs, which have had harmful effects, and which are exemplified by the Bhopal accident (UNEP 1987, 17).

Miscarriages and Birth Defects

A number of reports have disclosed the effects of contaminants on the developing foetus (teratogenic damage). One illustrative case occurred in Japan, where a chemical company discharged industrial wastes containing poisonous mercury into the Minamata river. This dumping, dating back to the 1950s, caused the so called

---

[1]A variety of other factors, of which many are directly related to the degree of toxicity, are also of importance and are described in more detail elsewhere (Postel and Heise 1988)

Minamata Disease. The mercury contained in the wastes entered the food chain through the water and fish, which was consumed by local inhabitants. As a result, mothers who had eaten a great deal of the fish and shellfish of Minamata Bay during their pregnancy bore children with "serious mental retardation, disturbance of growth...loss of consciousness and deformity of limbs," besides having a number of other adverse health effects (Manz 1989, 62). A similar hazardous waste dump occurred in Indonesia, causing "child and infant deaths" (Manz 1989, 63). In sum, teratogenic effects can be hidden, and they range from reduced fertility to total sterility, to miscarriages and still born or deformed babies (Gusman 1982, 50).

Long Term Damage

Aside from the health risks described above, the uncontrolled dumping of hazardous waste creates a potential for long-term damage, such as genetic changes and harm to the immune system. Naturally, long-term damage is the most difficult to ascertain, and very little research has been undertaken on the health of people who have been exposed to low concentrations of toxic substances over long periods (20 to 30 years) (UNEP 1986, 93). The extent of long-term effects depends on various factors, such as different ecological and weather conditions in LLDCs and NICs, nutritional aspects, cultural differences, and socioeconomic conditions. These factors may also influence the probability and effects of exposure to toxic substances (Yakowitz 1989, 2).

The long-term risks of waste dumping can be compared to the better-known impact of use, misuse, and overuse of pesticides in LLDCs and NICs. As has already been proven, toxic substances such as DDT, appear in foods grown far away from the location where the pesticides were originally used (Weir and Shapiro 1981). All indications are that the spread of harmful substances through the eco-system, due to improper disposal of hazardous wastes, will have equally catastrophic long-term implications. It should be kept in mind, in this context, that some materials might be very resistant and therefore remain hazardous for a long time. Their detoxification depends on their vulnerability to various break-through mechanisms: microbiological, photochemical, oxidation/reduction, their volatility, their solubility in water and so forth (WHO 1989, 104).

SUMMARY

Uncontrolled disposal of hazardous wastes can cause fires, explosions, and air, water, and soil pollution, contaminate food and drinking water, and directly or indirectly harm people. Altogether, the list of what can happen from waste dumps is quite alarming, and, in practice, "most of the things that could go wrong have indeed occurred. In fact, the incidents that have hit the headlines are probably only a few of those that have actually taken place: many more are likely to have gone unreported" (UNEP 1986, 115–116).

A study by UNEP and the World Health Organization (WHO) on contamination of water, soil, and air revealed that "the degree of contamination is worse in LLDCs and NICs than it is in most of the developed ones" (*Chemical & Engineering News* 1988, 8). Contamination causes environmental degradation, which has its direct consequences for people and land in LLDCs and NICs: it often results in the dislocation of people, a break-down of community infrastructures, the loss of productive land, and of economic activities (Nicholson-Brown 1986, 51). The combination of these effects has had a severe consequence, namely the creation of *environmental* refugees. Today, there are approximately 500 million environmental refugees—the single largest class of displaced people in the world, according to the United Nations Environment Programme (Keller 1988; Jacobson 1988; Jacobson 1989, 75). A majority of these refugees live in the developing world. A hazardous waste policy that allows exports to these parts of the world must carefully weigh its environmental consequences to these vulnerable people.

HIGH ECONOMIC COSTS

The export of hazardous wastes to other countries, whether industrialized or less developed, can also result in unanticipated high economic costs. These can be incurred from improper disposal, which will require cleanup of the site, final disposal later, compensation for poisoned victims, and potential loss of resources. The extent to which these costs occur depends on factors such as regulatory standards for disposal, legal provision for liability, and others. As argued in Chapter 3, the combined result of these factors in industrialized countries was an increase in disposal costs, which then became the principal economic argument to pursue waste export.

The following sections briefly discuss potential economic losses, if imported wastes are not managed in an environmentally sound manner.

Costs of Cleanup and Final Disposal

The dumping of hazardous wastes in landfills, without appropriate preparation of the landfill site and follow-up monitoring, generates an acute risk of environmental damage, resulting in large financial burdens to communities and governments. Many industrialized countries that allowed such dumps in the past are experiencing environmental degradation and are facing substantial expenses for cleanup and final disposal. In the United States, the *Superfund* program, designed to clean up old waste dumps, has already reached a budget of US\$ 10 billion, but 10 years ago, the amount needed to clean up all contaminated sites

was already estimated at US\$ 28–55 billion, by the President's Council on Environmental Quality (Third World Network 1988). Today, this amount has certainly increased, not only due to inflation, but also because disposal costs for incineration have exponentially risen (see Chapter 3). In other industrialized countries, the circumstances are similar (i.e., in West Germany, the costs for the cleanup of the 35,000 dump sites are estimated to be between US\$ 9–27 billion) (Evangelischer Pressedienst 1988).

The decision whether to dispose of the wastes in an environmentally sound manner or simply dump them depends, above all, on the laws regulating the management of hazardous wastes, as well as the enforcement capacity of the relevant authority. In a firm's financial analysis, the rate of discount and the marginal increase of disposal costs will play an important role in that decision. Potential future liability may be a factor encouraging immediate disposal; the development of new waste management technology may encourage temporary storage. The uncertainty of these and several other factors makes it difficult and unpredictable for a firm to fully assess future disposal costs. A firm may also be guided by the fact that it might not have to bear future costs, as it may be out of business. Empirical evidence seems to suggest that in the long term it is economically more beneficial to dispose of the wastes in an environmentally sound manner than to dump them and possibly be required to clean up later (World Bank 1989, 3).

The full extent of the financial burden resulting from the cleanup of dump sites in LLDCs and NICs cannot yet be fully assessed, as it is not clear what international obligations will exist concerning hazardous waste exports.[2] Certainly, illegal dumping is unacceptable. When discovered, it is also far costlier than correct disposal, as amply demonstrated in the Italo-Nigerian waste deal examined in Chapter 2 (Case 3). Because the ship employed to carry the unwanted wastes back was unable to unload its cargo in six countries, it landed finally in Italy, where the Italian Ministry of the Environment estimated the total costs for the operation to be US\$ 14.3 million for the first 2,000 tons of wastes (Greenpeace 1988, 7). Appropriate disposal of this 2,000 tons of waste could have cost up to a maximum of US\$ 4,399 per ton (1989 dollars),[3] or some US\$ 8.8 million. Given the current price for incineration, the Italian waste project could therefore have saved some US\$ 5.5 million, if the wastes had been incinerated as a first step.

---

[2]A rule adopted by the OECD "holds the originating nation responsible for waste that is disposed of improperly" (*New York Times* October 16, 1988).

[3]The *highest* price for PBC-incineration that could be obtained was US\$ 3,305 per ton, in 1986 dollars. This base price calculated for 1989 dollars would be US\$ 4,399 (U.S. EPA. Office of Policy Analysis 1985, 3–19).

Treatment and Compensation for Poisoned Victims

The hazards of uncontrolled dumps in LLDCs and NICs to human health and life incur not only the expenses for treatment and cure, but may also culminate in high financial costs for compensating contaminated victims. As recent public stands taken by some of these countries have shown, most LLDCs and NICs will no longer accept export practices as they were conducted in the past.

To date, there is no information available on whether or how much money was paid in compensation to any victims of hazardous waste export schemes. There are wide differences in the export contracts, and legal mechanisms for imposing liability are often weak or absent. The lack of an international system of accountability for improper disposal abroad allows hazardous waste generators to escape liability and thus externalize the costs of their careless activities (Handley 1989, 10172). Some of the past contracts were entirely illegal, since they involved bribery or falsification of documents. Other contracts may have been valid, but the importing country did not have any or not enough information to make a sound decision. As a result, an acceptable policy on hazardous waste exports must clarify this issue and include liability provisions for waste generators, brokers, transporters, and recipient countries.

Corporations in industrialized countries may, however, still find it more attractive to export wastes abroad and accept having to pay potential compensation in the future. Their economic analysis is most likely that damage costs in LLDCs and NICs are lower than in industrialized countries, which can be concluded from the Bhopal accident.

Destruction of Resources

The destruction of valuable natural resources represents another potential economic loss for countries that import hazardous wastes. As described above, the unloading and disposal of the wastes mainly occur along water systems. In the Nigeria waste scheme, the dumpsite was connected to the Benin river, which irrigates the farmland of 30,000 farmers. As a result of this one waste dump, the government felt forced to request the local people to stop harvesting their farmland, to prevent the intake of unsafe amounts of chemicals (Ogunseitan 1988, 15). The economic loss is thus not limited to the resource water, but can, in fact, be an entire harvest or more.

To estimate the potential damage is very difficult, given the scantiness of data and the uncertainty of numerous factors that play a role in such an assessment. Furthermore, existing methodologies of economic analysis are limited in evaluating the possible damage or destruction of marine life and the bio-diversity of tropical forests. The fact is, however, that environmental disasters such as the

Exxon Valdez accident in Alaska (USA) can cause a magnitude and diversity of damage that may indeed be irreparable.

The evaluation of the most important economic consequences of hazardous waste imports for LLDCs and NICs suggests that these imports make no sense economically, "except in terms of the most primitive short-term financial analysis," as was summarized by an OECD expert (Yakowitz 1989, 16). He went on to point out that improper disposal can end up "costing a society 100 to 1000 times more than environmental [*sic*] sound management" of the wastes (Yakowitz 1989, 16).

POLITICAL AND DIPLOMATIC DIFFICULTIES

Many of the past waste exports from industrialized countries to LLDCs and NICs were "sham recycling" operations or involved some other unethical and illegal actions, such as bribery and forgery of documents. In some cases, even high officials of importing and exporting governments, and diplomatic missions were involved in the operations. In several cases, when the waste schemes became public, they caused internal investigations within governments. Others harmed diplomatic relations between the governments involved.

The issue of hazardous waste exports raises moral questions about the ethics of exporting risks and hazards abroad. Consequently, the debate after the first waste dumps became publicly known was very emotional. It literally touched the consciousness of LLDCs and NICs, many of which saw in this practice a "new form of colonialism," or "garbage imperialism." In order to understand this emotional debate, one can look, for example, at the sociopolitical context in much of Africa, where many countries reached political independence in the post–World War II period, but since then have been confronted with tremendous economic difficulties. Today, these countries feel they operate in a "world in which the resources gap between most developing and industrial nations is widening, in which the industrial world dominates in the rule-making of some key international bodies, and in which the industrial world has already used much of the planet's ecological capital," as the Brundtland report stated (WCED 1987, 5). As a result, they are very sensitive towards issues involving possible damage to their country.

At one point in the late 1980s, as a reaction to hazardous waste scandals, the governments of Venezuela and Nigeria recalled their ambassadors to Rome for consultations, and Nigeria indicated that the matter would be brought to the International Court of Justice at The Hague. The government of Guinea-Bissau arrested Norway's Honorary Consul-General, who was the director of a joint company that imported waste into Guinea-Bissau. Following these diplomatic irruptions, the issue of hazardous waste exports was addressed in all major international organizations by delegations from LLDCs and NICs. Several organizations, such as the Organization of African Unity (OAU), the Economic Organization of Western African States (ECOWAS), the African, Caribbean and Pacific States (ACP), and others condemned waste exports and called for a total ban. One industrial country was also prepared to ban waste exports to developing countries: the Netherland's environment minister called for an export ban in the EEC, but was successfully opposed by Britain's environment minister (Islam and Smit 1988).

On a national level, over 40 countries have enacted legislation prohibiting toxic waste import. In addition, a number of African countries, namely Gambia, Guinea, Liberia, Togo, and the Ivory Coast, have adopted heavy penalties—fines, jail terms, and payment of cleanup costs—for persons convicted of importing hazardous wastes illegally. Ghana has formed a "Toxic Task Force" to investigate waste schemes, and Nigeria has debated a decree for a death penalty for anyone convicted of importing toxic waste (New York Times September 22, 1988).

In general, the industrialized countries do not seem to want to damage foreign relations either. In fact, the possibility that a hazardous waste-related incident might occur worries officials in these countries. Senator George Mitchell (D-Me., USA) said in a 1984 hearing, "If I were the U.S. Secretary of State, I would want to be sure that no American ally or trading partner was saddled with U.S. wastes it does not want or does not have the capacity to handle in an environmental sound manner." (Porterfield and Weir 1987, 344). This Senator is obviously concerned by the potential worsening of relations and therefore views waste exports with considerable caution, although it is not clear who is meant by *American ally or trading partner.*

The export and improper disposal of hazardous wastes bears other political consequences that become visible in the long term. Dumped wastes may reduce the future development potential of a country. Depressed local economies face rising poverty and unemployment, which, through consequent overexploitation, improper use, and so forth, place increased pressure on environmental resources. Moreover, due to their increasing economic problems, many governments have cut back on efforts to protect the environment. The result is a deepening and widening environmental crisis that can lead to severe pressures on political systems and leaders (WCED 1987, 6-7; Nicholson-Brown 1986). The political stability in LLDCs and NICs is thus directly related not only to economic conditions, but increasingly to environmental factors. Environment and development are becoming increasingly interdependent, and a policy of hazardous waste export-import must consider this reality.

A last word on political consequences: the general development of waste-importing countries, particularly through wider and improved education, better information flows, and more general awareness, may lead to popular resentment of waste imports and thus damage relations between countries. Without doubt, the exporting countries run the risk of being charged with taking advantage of unequal relations, "neo-colonialism," and exploitation (Yakowitz 1989, 15).

DIMINISHED PRESSURE FOR WASTE REDUCTION IN INDUSTRIALIZED COUNTRIES

Many industrialized countries have encountered increasing problems in managing their hazardous wastes. As explained in detail in Chapter 3, the disposal costs of hazardous wastes have continuously increased in the past decade. One of the approaches to solve the *waste crisis* in individual countries was to export abroad. This option, however, has direct consequences on efforts to minimize wastes in industrialized countries.

Waste minimization is the reduction of hazardous wastes that are generated, treated, stored, or disposed of. It includes any source reduction or recycling activity undertaken by a generator that results in either (1) a decrease in the total volume or quantity of hazardous wastes, or (2) the reduction of toxicity of hazardous wastes, or both (World Bank 1989, 164). Besides recycling and treatment, source reduction is the most crucial component of waste minimization, consisting of product substitution and source control.

Source control involves a number of activities in the selection of materials, technologies, and procedures. All of these activities are related and affected by economic and legislative factors. Legislative initiatives to create financial incentives for waste generators and regulations ensuring the environmentally-sound disposal of the wastes are the two main control instruments. Stricter regulations as to how certain wastes must be disposed of have indeed led to measures to minimize waste. An OECD report concludes that, as a result of governmental regulations within the OECD, the volumes of hazardous wastes could be reduced by 15 percent (Roelants du Vivier 1988, 78). However, as long as companies have access to sea incineration, sea dumping, and the uncontrolled export of hazardous wastes, there is little incentive for recycling or source reduction. As international efforts gain momentum to limit or ban the option of sea disposal, waste exports become even more attractive. The waste minimization strategies that have been adopted by industrialized countries over the past few years may be seriously undermined if the option of unlimited and uncontrolled waste exports is available (Roelants du Vivier 1988, 80).

Currently, there are legislative controls in some countries allowing the export of hazardous wastes only when it is proven that no technically or economically feasible disposal option for a certain waste exists in the country of generation (Roelants du Vivier 1988, 125). In practice, the term "technically or economically feasible" has represented the loophole for waste exports, while providing little incentive to search for new solutions. In the future, however, the potential long-term harm that could result even to industrialized countries, not to mention the responsibility carried for waste-generating activities, should be reason enough to find environmentally and economically acceptable ways of managing the waste within the generating industries, communities, and countries (Yakowitz 1989, 16).

References

Blackburn, A.M. ed. 1986. Examples of Responses. *Pieces of the Global Puzzle*. Golden: Fulcrum Inc.

Chemical & Engineering News. October 3, 1988. Third world has most chemical contamination. *Chemical & Engineering News*, pp. 8–9.

Clark, R. and J. Palmer. 1983. *The Human Environment: Action or Disaster*. ed. R. Lumb.

Consumers' Association of Penang. 1989. What is hazardous waste. In *Toxic Terror: Dumping of Hazardous Wastes in the Third World*, pp. 98–101. Penang, Malaysia: Third World Network.

Evangelischer Pressedienst. July 1988. EPD-Entwicklungspolitik. *Evangelischer Pressedienst*, FRG. Nr. 14/15.

Greenpeace. 1988. Regional Updates on International Waste Trade Schemes. *Greenpeace Waste Trade Update* 2(1). Washinton, D.C.: Greenpeace.

Gusman, S. et al. 1989. *Die Kontrolle von Umweltchemikalien—Nationale und Internationale Fragen*. Berlin.

Handley, J. 1989. Hazardous waste exports: a leak in the system of international legal controls. *Environmental Law Reporter* 84(4):10171–182.

Islam, S. and P. Smit. August 1988. Dirty Games in Brussels. *South*, pp. 37–41.

Jacobson, J.L. 1988. *Environmental Refugees: A Yardstick of Habitability*. Washington, D.C.: Worldwatch Institute.

Jacobson, J.L. 1989. Abandoning homelands. *State of the World* 1989. ed. Worldwatch Institute, pp. 67–75. New York: Norton & Company.

Keller, U. 1988. Globale Bedrohung. *Dritte Welt Presse* 1(5):1–2.

Manz, G.M. 1989. Mercury poisons people of Minamata Bay. In *Toxic Terror: Dumping of Hazardous Wastes in the Third World*, pp. 60–63. Penang, Malaysia: Third World Network.

New York Times. September 22, 1988. African nations barring toxic waste. *New York Times*.

New York Times. October 16, 1988. Europe's failing effort to exile toxic trash. *New York Times*.

Nicholson-Brown, J.M. 1986. Converging worlds: environmental events and the free market and policy development. In *Pieces of the Global Puzzle*, ed. A. Blackburne. Golden: Fulcrum.

Ogunseitan, S. November 1988. Nigeria: the drums are gone but the poison remains. *Panascope* 9:15–16.

Porterfield, A. and Weir. October 3, 1987. The export of U.S. toxic wastes. *The Nation*, pp. 341–44.

Postel, S. 1988. Controlling toxic chemicals. *State of the World* 1988, ed. Worldwatch Institute, pp. 119–136. New York: Norton & Company.

Postel, S. and L. Heise. 1988. *State of the World* 1988, Worldwatch Institute. New York: Norton & Company.

Roelants du Vivier, F. *Les Vaisseaux du Poison: La Route des Dechets Toxiques*. Paris: Sang de la Terre.

Third World Network. 1989. Toxic waste dumping in third world countries. In *Toxic Terror: Dumping of Hazardous Wastes in the Third World*, pp. 8–25. Penang, Malaysia: Third World Network.

Todd, H. 1989. Impact of toxic wastes in asia. In *Toxic Terror: Dumping of Hazardous Wastes in the Third World*, pp. 53–57. Penang, Malaysia: Third World Network.

United Nations Environment Programme. 1983. *The State of the Environment 1983*. Nairobi: UNEP.

United Nations Environment Programme. 1986. *The State of the Environment: Environment and Health*. Nairobi: UNEP.

United Nations Environment Program. 1987. *Report of the African Ministerial Conference on the Environment on the Work of its Second Session*. African Ministerial Conference on the Environment (AMCEN). UNEP/AEC. 2/3. July 20, 1987. Nairobi: UNEP.

U.S. EPA. Office of Policy Analysis. 1985 *Survey of Selected Firms in the Commercial Hazardous Waste Management Industry*. Final Report. November 1986. Washington, D.C.: U.S. EPA.

U.S. News & World Report. November 21, 1988. Dirty jobs, sweet profits. *U.S. News & World Report*.

Yakowitz, H. 1989. Monitoring and control of transfrontier movements of hazardous wastes: an international overview, *OECD-Technical Paper*. Paris: OECD.

Weir, D. and M. Schapiro, 1981. *The Circle of Poison*. San Francisco: Institute for Food and Development Policy.

Wesley, M. 1971. *Man and His Environment: Waste*.

World Bank. 1989. The safe disposal of hazardous wastes: the special needs and problems of developing countries. *World Bank Technical Working Paper No. 93. Vol. I*. Washington, D.C.: World Bank.

World Commission on Environment and Development (WCED). 1987. *Our Common Future*. Oxford: Oxford University Press.

World Health Organization. 1989. Examples of toxic waste. In *Toxic Terror: Dumping of Hazardous Wastes in the Third World*, pp. 129–132. Penang, Malaysia: Third World Network.

Part II

Evaluation of National, Regional, and International Policies

With exports of hazardous wastes, leaders in industrialized countries have become increasingly concerned about possible adverse effects. As a result, a number of industrialized countries realized the need to regulate the exports of hazardous wastes. In the remainder of this book, I discuss policies and initiatives on the national, regional, and international levels, to control movements of hazardous wastes across borders and analyze relevant policy options.

The analysis below examines various operational mechanisms and criteria that are considered essential for a sound hazardous waste export policy. I will examine the framework of existing national hazardous waste policies, the form of a manifest system and procedures for notification and prior informed consent, and the existence of mechanisms for effectively implementing, enforcing, complying with, and monitoring the provisions. The examination also includes the consideration of damage caused by toxic releases and analyzes liability and insurance provisions. Information exchange, technology transfer, and so on, are important mechanisms for ensuring the policy's operational realization and thus will be assessed. The analysis will show if and how each policy meets the broader goals of efficiency, risk minimization, equity, and sustainable development.

Various policy options are discussed in the following three sections of this book:

1. A policy of continued current practice;
2. Unilateral restrictions; and
3. International policies that are further divided in bilateral agreements, global ban, regional agreements, and other policies.

5

Continue Current Practice

Proper hazardous waste management (HWM) requires a legal, technical, political, and economic infrastructure capable of supporting regulatory provisions, technical standards and monitoring, and adequate financial resources. Currently, there is a wide gap in these infrastructures between industrial and low and least developed countries (LLDCs) and newly industrialized countries (NICs). The social and cultural attitudes and economic status of citizens of different countries also vary widely. In industrial countries, environmental organizations have played an increasingly important role in environmental policy formulation and the strengthening of hazardous waste regulations. In contrast, countries in Eastern Europe, as well as LLDCs and NICs, confronted with very limited resources, have given priority to economic development, while environmental protection has lacked the necessary political and economic support. As a result, there is no uniformity in environmental standards, pollution control, and environmental protection on a global level.

In each country, the particular hazardous waste policy and the export practice stems from these differing economic-political priorities and sociocultural values. Furthermore, the consequences of waste export must be interpreted in the same context. Although the boundaries are often blurred, most countries can be placed into one of two groups: industrialized countries on the one hand, and LLDCs and NICs on the other hand. This categorization is broad, and there are significant differences within each group. However, hazardous wastes are almost exclusively exported from industrialized countries and imported to LLDCs and NICs. This section describes and analyzes the main issues concerning hazardous waste exports in industrialized countries, LLDCs, and NICs. It also takes a look at Eastern Europe and discusses the role of Export Processing Zones (EPZ) and Transnational Corporations (TNC), which play an important role in the context of economic development and environmental protection in LLDCs and NICs.

INDUSTRIAL NATIONS

Industrialized countries, which have enormous capacity to produce and develop numerous chemical substances and are generally aware of the potential hazards resulting from their improper handling or disposal, have developed systems of hazardous waste management in order to control and minimize the risks involved. A proper system includes environmental impact assessment (EIA) of existing and projected industrial projects and requires regulations and legislation covering standards for generation, transport, storage, and disposal of hazardous wastes. In addition, enforceability of provisions and an effective monitoring system to ensure correct conduct and the tracing of wastes is also necessary. Finally, provisions for liability and mandatory insurance should require the compensation of personal and property damage and reimbursement for cleanup in case of an accident.

Hazardous Waste Regulation

The objective of hazardous waste legislation is to provide a comprehensive system of hazardous waste management from the time of the manufacture of the chemical substance and its use to the ultimate disposal of the hazardous wastes produced in the process. Such a system is called "cradle to grave." In the process of developing such a system, many countries have established specialized agencies, such as the Environmental Protection Agency in the United States or a "Ministry for the Environment," as in several European countries. Specific legislation for hazardous waste management was adopted by almost all of the OECD member countries by 1985.[1] A summary of legislation on hazardous wastes by the OECD in 1985 shows that out of 24 member states, only Australia, Greece, Iceland, New Zealand, Portugal, and Turkey had no regulations on hazardous waste management. According to this list, two EEC member states—Greece and Portugal—have no laws regulating hazardous wastes. Although the Single European Act requires its member states to harmonize their legislation, including environmental legislation, so far fundamental difficulties have emerged from harmonizing definitions and standards (Laurence, Wynne 1989). In the EEC, in which hazardous waste laws among member states are still very diverse, some member states tightened their laws, and others did not, giving additional incentives to export hazardous wastes to countries that have less stringent laws or that lack adequate laws altogether (Du Vivier 1988).

Thus, higher regulatory standards and more stringent laws, coupled with high public awareness, has affected the industry's practice in dealing with hazardous

---

[1]For an analysis of hazardous waste management in selected industrialized countries, see OECD. 1985. *Transboundary Movements of Hazardous Wastes: Legal and Institutional Aspects.* Paris: OECD

wastes. Facing public pressure and less disposal capacity, the industry increasingly moved to export wastes abroad.

Regulation of Transboundary Movements of Waste

For industrial countries, the expansion of national hazardous waste legislation to cover their transfrontier movements appears to be a consistent process, given the potential risks of accidents and spills. Without any specific national legislation, transboundary movements of hazardous wastes were regulated by international agreements covering standards for the safe carriage of hazardous wastes or dangerous goods (Hannequart 1985). Up until 1985, several industrial countries, particularly OECD countries, promulgated specific legislation concerning transboundary movements of hazardous wastes.[2] These legislative efforts can be attributed largely to activities within the OECD. Based on a country's specific needs, legislation was specifically enacted for combinations of hazardous waste imports, exports, and transit. Table 5-1 shows what types of legal provisions were enacted by 16 industrialized countries. As can be seen from the table, many industrialized countries regulated hazardous waste imports, and only three countries did not regulate exports at all. The transit of hazardous wastes appears to be of least importance. Only two countries had such provisions, while another two were in the process of proposing them. The specific types of control exercised through the provisions differ. Most often, they require information on the wastes from either the exporting or importing country, as well as authorization of exports by the issuing of permits to carry out waste shipments.[3]

The existing provisions also designate different responsibilities to government. Under most pre-1985 legislative schemes, the responsibility for an entire waste export scheme did not rest with one agency, but was shared among several independent authorities. Governmental authorities had mainly the responsibility to receive and send notifications, and to issue licences to exporters and importers. The responsibility of the customs service was to carry out inspections on the waste cargo and to control the trip ticket or manifest. Usually, a different authority was responsible for securing public health concerning hazardous wastes, by checking security and emergency measures (Hannequart 1985). Effective enforcement thus required efficient communication and coordination among the different authorities. Unfortunately, this has often not been the case. Table 5-2 identifies the responsibilities of different authorities for transboundary waste movements in federal, state, departmental, and local governments.

---

[2]For a discussion of the U.S. policy, see Chapter 6.

[3]As indicated above, a summary and overview of national laws on transfrontier movements of hazardous wastes was published (OECD 1985). However, some of these laws are outdated by now.

TABLE 5-1.   Specific Legal Provisions Relating to Hazardous Wastes Export, Import, or Transit.

	Specific Legal Provisions Relating to:		
Country	Imports of Hazardous Wastes	Exports of Hazardous Wastes	Transit by Hazardous Wastes
Germany	yes	proposed	proposed
Austria	yes	no	no
Belgium	yes	yes	no
Canada	proposed	proposed	no
Denmark	yes	no	no
United States	no	yes	no
Finland	yes	yes	no
France	yes	no	yes
Ireland	yes	yes	no
Italy	yes	yes	no
Luxemburg	yes	yes	yes
Norway	proposed	proposed	no
Netherlands	yes	yes	no
United Kingdom	yes	yes	no
Sweden	proposed	yes	no
Switzerland	yes	proposed	proposed

Source: OECD. 1985. Transfrontier Movements of Hazardous Wastes: Legal and Institutional Apects, Paris: OECD.

Evaluation

Most imports and exports of hazardous wastes are within industrialized countries and appear to represent a conventional economic activity (NYT 23.3.89, OECD 1985). Industrialized countries presently feel compelled to export part of their hazardous wastes, mainly because they do not have the necessary disposal capacity or at least not at attractive prices, but also because of more stringent regulatory standards and public pressure. Given their common market structure, similar environmental regulations, and institutional capability to manage hazardous wastes, transboundary movements of such wastes could be considered a trade among equal partners. This trade sometimes produces economic benefits. Great Britain, one of the major importers of hazardous wastes in Western Europe, was able to make good financial profits from the disposal business (Frey 1989). It has also been suggested that it is likely that these exports may also have resulted in environmental benefits, "because the waste is disposed of more completely and safely within the country of import than would have occurred in the country of export" (Frey 1989, 509). Besides these benefits, industrialized countries want to remain flexible in their hazardous waste management policies and therefore lobby for a policy to continue transboundary waste movements.

TABLE 5-2.   Basic Government Responsibilities in Regard to Transfrontier Movements of Hazardous Wastes (in 1985).

OECD Member Countries	Specific Legislation on Hazardous Wastes
1. Germany	Act of June 7, 1972, amended interalia June 21, 1976. Order of 1974, 1977, 1978, etc.
2. Australia	—
3. Austria	Act No. 186 of March 2, 1983.
4. Belgium	Act of July 22, 1974; Royal Decrees of 1976; Regional Decree of 1983.
5. Canada	Act of 1980 on transport of hazardous material; regulation of December 19, 1982.
6. Demark	Act of May 24, 1972; Decree No. 121 of March 17, 1976 and No. 323 of July 3, 1980; Act of May 18, 1983; Act of 1973 on the environment.
7. Spain	Catalonian Act of March 24, 1983.
8. United States	Act 94.580 of October 21, 1976, Regulation of 1980, 1981, etc.
9. Finland	Act No. 673 of August 31, 1978 amended on February 30, 1981; Order of July 5, 1983
10. France	Act of July 15, 1975; Decree of August 19, 1977; Order of July 5, 1983.
11. Greece	—
12. Ireland	Regulation No. 33 of 1982.
13. Iceland	—
14. Italy	Act No. 915 of September 10, 1982.
15. Japan	Act No. 137 of 1970, amended in 1976; Decree No. 300 of 1971.
16. Luxemburg	Regulation of June 18, 1982.
17. Norway	Act No. 6 of March 13, 1981.
18. New Zealand	—
19. Netherlands	Act of February 11, 1976, Decree of 1977, 1980, etc.
20. Portugal	—
21. United Kingdom	Regulation No. 1709 of 1980.
22. Sweden	Òrdonnance No. 346 of May 22, 1975.
23. Switerland	1983 Environment Act.
24. Turkey	—

Source: OECD. 1985. Transfrontier Movements of Hazardous Wastes: Legal and Institutional Apects, Paris: OECD.

At the same time, industrialized countries recognize the need for regulatory control and monitoring in order to prevent environmental damage. This led to the promulgation of national laws regarding hazardous waste movements. Although the regulatory framework within industrialized countries varies, the overall terms of trade and their relative equal technical capability contributes to safe waste disposal to the extent that it is required in existing regulations and standards in industrial countries. Presuming equal standards for waste management, it is argued that this trade has not undermined national policies to minimize waste. However, a free trade within the EC would certainly undermine long-term national efforts for waste minimization and source reduction, as the pressure resulting from environmentally sound domestic waste management is lessened.

Recent discussions in the European Parliament and the Commission of the European Communities indicate that a trade in wastes even within the EEC is not considered an optimal solution. As a minimum, a uniform EEC-wide approach to the control of transfrontier movements of hazardous wastes is necessary. National approaches are not sufficient because nation-states have limited control over transnational issues, and the potential environmental hazards from hazardous wastes are not limited to one country. This has become clear from various activities in the EEC and OECD to design control instruments. An international approach is also required in order to harmonize hazardous waste standards and definitions, a prerequisite for an effective hazardous waste policy.

The increasing export of waste from industrialized countries to LLDCs and NICs has not only impeded waste reduction policies in industrialized countries, but also caused a variety of difficulties, as well as resulted in environmental damage. The terms of trade in these export schemes were distorted in that many LLDCs and NICs are heavily indebted, and currently have neither the capacity to make an informed decision on waste imports nor the necessary infrastructure to ensure environmentally sound waste disposal. Therefore, their willingness to import hazardous wastes may reflect indirect economic-political foreign pressure, rather than a domestically developed hazardous waste management policy. Moreover, the imported wastes are not managed in an environmentally sound manner, which should be the goal of any hazardous waste management policy. The particular circumstances and conditions of LLDCs and NICs will be discussed in the next section.

LESS INDUSTRIALIZED COUNTRIES

LLDCs and NICs face very different problems than do industrial countries. They have hardly any environmental groups exercising political pressure to promote environmental legislation. In the past, environmental issues were not a priority concern for LLDCs and NICs. Overwhelmingly, their political priority primarily has been and still is economic development "even at great cost to the environment" (Wolff 1972,7). Environmental issues have come on the agenda in recent years, only as increasing hazards to the environment began to threaten human existence in many LLDCs and NICs (WCED 1987). International and regional organizations played an important role in raising environmental issues. For example, the United Nations Economic Commission for Africa (UNECA) has set a twofold goal: to set priorities in environmental issues and to establish environmental institutions all over Africa (Tandap 1989). Furthermore, the environmental agenda of LLDCs and NICs is distinct from those of industrial countries. A 1972 report of the United Nations states:

> The major environmental problems of developing countries are essentially of a differ-ent kind. They are predominantly problems that reflect the poverty and very lack of de-velopment of their societies. They are problems...of both urban and rural poverty. In both the towns and the countryside, not merely the 'quality of life' but life itself is en-dangered...These are problems...which affect the greater mass of mankind (UN 1971).

The 1989 conference in Addis Ababa, Ethiopia, of the Council of Ministers of the Organization of African Unity included 7 topics relating to the environment, out of 23 concerning political, economic, educational, scientific, and cultural matters. Environmental issues dealt with included drought and famine, control of locusts and other migratory pests, development of an hydrological map for Africa, dumping of nuclear and hazardous wastes, report on refugees, maritime transport, and report of the African ministers of environment (OAU 1989). The list indicates that environmental matters are of crucial importance to LLDCs and NICs. How-ever, the topics themselves are quite different from the environmental issues of industrial countries. According to the list, only *dumping of nuclear and hazardous wastes* and perhaps *maritime transport* are issues of common concern. Recognition of these differences in environmental priorities are crucial for any multilateral negotiation between industrialized countries and LLDCs and NICs. They are particularly important in environmental negotiations in order to respond to the particular concerns of each party.

The State of the Environment in Developing Countries

Environmental pollution is worse in many LLDCs and NICs than it is in indus-trialized countries (O'Sullivan 1988). A study in Egypt, for example, reveals serious contamination of lakes from pesticides or discharge of mercury (Hamza 1983). By raising the annual per capita GNP and increasing population, industri-alization has generally had a higher priority than the reduction of pollution. As a result, water pollution caused by organic and inorganic wastes from industrial production is widespread and is considered one of the gravest environmental problems in LLDCs and NICs (Maurits la Riviere 1989). Developing and newly industrializing countries are facing environmental degradation in various forms.[4] Based on a particular country's industry, LLDCs and NICs generate hazardous

---

[4]Many of these are discussed in detail by the annual reports of the Worldwatch Institute's *State of the World* and UNEP's *State of the Environment.*

wastes mainly from the production of pesticides, dyes and pigments, organic chemicals, and fertilizers. In many of these countries, river pollution increases while efforts of decontamination are often neglected (Maurits la Riviere 1989).

Moreover, hazardous waste generation has increased with greater industrialization in some countries, leaving many developing and newly industrializing countries with much larger volumes of waste than in previous years (Yakowitz 1989). Although there is no systematic empirical estimate on the volumes of hazardous wastes generated each year, projections of hazardous wastes generation in developing and newly industrializing countries are quickly increasing. Newly industrializing countries, like Thailand, are experiencing particularly serious hazardous waste problems and are seeking international assistance to solve them.[5]

Constraints in Hazardous Waste Management

In LLDCs and NICs, the possibilities to manage hazardous wastes in an environmentally sound manner are limited and constrained by several problems. LLDCs and NICs may not be prepared to manage foreign waste. This was well summarized in a speech by the Nigerian Ambassador to Vienna before the Board of Governors of the International Atomic Energy Agency (IAEA). He stated that the import of highly toxic waste to LLDCs and NICs is not acceptable when this country:

> ...is not informed on and lacks the experience and expertise in such basic matters as the siting of geological repositories, waste packaging, repository design, operation, shut-down and surveillance. In such a situation the country cannot be in a position to conduct independently, safety assessment and monitoring of geological repository (Mgbokwere 1988).

Specifically, LLDCs and NICs are constrained by the lack of a regulatory framework, lack of infrastructure and limited financial resources, and various sociopolitical factors.

Lack of a Regulatory Framework
Currently, few LLDCs and NICs have legislation covering the generation, storage, transport, or disposal of hazardous wastes. The lack of a regulatory framework to control the generation of hazardous wastes and their disposal in LLDCs and NICs has often led to a practice of uncontrolled landfilling or even indiscriminately dumping the wastes into the environment (Postel 1988). Accentuating the problem

---

[5]In Thailand, for example, a recycling project for waste lubrication oil is being set up with assistance from the United Nations Industrial Development Organization in order to collect and manage these wastes in an environmentally sound manner. In the past, large quantities of these wastes were uncontrolled.

is a lack of technological sophistication to determine the identity and toxicity of substances in wastes. As a result, the promulgation of a list of substances considered as hazardous may have little practical effect. Similar problems arise from developing safe transport, storage, and disposal standards. With regard to waste imports, a general weakness stems from porous borders that are not controlled effectively. Even assuming probity among governmental officials and inspectors, it is extremely difficult to stop determined hazardous waste brokers from their activities as "relatively small pay-offs have enabled dumpers to flout customs procedures" (George 1988).

Furthermore, the simple adoption of a regulatory framework from industrialized countries would perhaps not have the desired effect because a variety of differences exist in institutional capability. Complex, lengthy, expensive, and time-consuming EIAs, as required in developed countries, are not realistic tools to assess the impacts of industrial projects of LLDCs and NICs with limited expertise, lack of an operational infrastructure, and limited resources. Such complex EIAs may, under certain circumstances, even hinder the process of overall development (Maltezou, Biswas, and Sutter 1989).

Another important problem results from the predominance of small-scale industry in LLDCs and NICs. In contrast to industrialized countries with large-scale industrial projects, which often have the technical expertise and financial resources, LLDCs and NICs have numerous small-scale plants generating small amounts of waste, each of which may not represent a significant hazard. However, the collective impact of a series of small plants on the environment can be significant, and in some cases even worse than a single large plant (Maltezou, Biswas, and Sutter 1989). Regulatory systems in industrialized countries often have excluded small-scale generation of hazardous wastes because of the relative insignificance of the overall volume thus generated.[6] In LLDCs and NICs, however, small volumes of waste are the dominant problem which needs immediate attention. Although the regulation of small volumes of waste is important and recognized, its management could represent a disproportionate financial burden to small firms. Often, small firms are economically weak and, if they want to remain in business, simply cannot afford expensive disposal for small but highly toxic amounts of wastes. A regulatory system has to acknowledge the industry structure in LLDCs and NICs and be fashioned accordingly.

Lack of Infrastructure and Limited Financial Resources
The lack of adequate waste management technologies, trained personnel, and institutional support represent practical constraints for industries in LLDCs and

---

[6]In the United States, under RCRA, the production of up to 100 kg/months and the storage of up to 1000 kg of hazardous waste is excluded from the manifest system. [42 U.S.C.A. §6921 (d)(3)].

NICs. Taiwan, a country with tremendous development over the past decade, describes its own environmental state in *Taiwan 2000*:

> Many of the industries that thrive on Taiwan…are prolific producers of hazardous, and sometimes very hazardous, substances. With few, if any, disposal facilities and no institutional mechanisms to ensure the safe disposal of such substances, it is suspected that large quantities of hazardous wastes could be dumped into rivers or onto the ground, or, at best, into rusting barrels.…Unfortunately, pollution control laboratories in Taiwan are ill prepared to monitor hazardous waste problems (Steering Committee Taipeh 1989, 8).

Since the financial resources of LLDCs and NICs are very limited, many countries find themselves trapped in a position of having a choice only to invest either in general development projects or in environmental pollution control. However, environmental protection and the general welfare of a society are increasingly interdependent. Thus, a decision to control environmental pollution and thereby contribute to environmental protection ultimately contributes to development. The sole promotion of industrial development may be detrimental in the long term. Nevertheless, financial constraints (i.e., limited foreign exchange and access to hard currency, which make it difficult to finance appropriate technologies) is another major reason why environmental protection has limited support in LLDCs and NICs (World Bank 1989).

Sociopolitical Factors
The lack of general awareness of environmental hazards from improper disposal of hazardous wastes caused by insufficient information and various other sociocultural reasons often lead to little public demand for action in LLDCs and NICs. The immediate needs and seemingly more urgent problems such as food, housing, and so forth, are the focus of political goals (World Bank 1989). Although some countries have developed a regulatory framework for hazardous waste management, often there are no incentives and will for industry to comply. Particularly important is the lack of enforcement. Administrations and institutions in LLDCs and NICs often do not have the resources and capability to enforce and monitor regulations (Mahajan 1989). Thus, in some LLDCs and NICs, the state apparatus is neither well developed nor strong in exercising authority to control and monitor hazardous waste disposal (Yakowitz 1989).

Conclusion

The majority of LLDCs and NICs are presently not capable of managing their own wastes in an environmentally sound way. In the future, LLDCs and NICs will have to fill a widening gap of environmental protection with regard to hazardous waste disposal and must ensure that they do not make the same mistakes industrialized countries made during their development process, when many wastes were recklessly dumped. In order to be better able to accomplish self-reliant waste management, there is a great need for increased education and training of personnel. Financial and institutional aid, as well as efficient transfer of waste disposal technologies are crucial factors as to whether LLDCs and NICs will be able to successfully tackle waste management.

Ironically, it could be argued that, in the short term and for environmental reasons, hazardous wastes should be exported to industrialized countries to ensure safe disposal. This idea emerges as a serious proposal stemming from existing environmental hazards, such as improper storage of pesticides that are unsafe to use and therefore need disposal.[7] However, since LLDCs and NICs can not finance such exports, such an option has not been taken seriously. In the long term, waste minimization and source reduction policies, including the substitution of less toxic or even nontoxic methods of pest control, are the prefered option in order to avoid waste-related problems altogether.

Given these conditions in less developed countries, the proposals to ensure the safe disposal of hazardous wastes exported from industrialized countries to LLDCs and NICs are at present unfeasible and unrealistic. Considering the lack of environmental protection infrastructure and regulation, an unrestricted export of hazardous wastes to LLDCs and NICs would shift the entire burden of regulation to governments that may not even be aware of the problem and are much less able to cope with inherent problems. If such transfrontier waste movements take place nevertheless, it must be assumed that most, if not all, wastes shipped to LLDCs and NICs are disposed of in uncontrolled landfills or dumped. The potential environmental effects of such a practice could be disastrous.

EASTERN EUROPE

The political changes throughout Eastern Europe have for the first time revealed the state of the environment in these countries. The centrally planned economies in Eastern Europe have not only resulted in catastrophic economic conditions but have also caused irreversible damage to the environment. The following overview will examine the conditions of hazardous waste management and the policies of waste imports and exports in selected eastern European countries. The basic reasons for the present critical situation of the environment lie inherently in the socioeconomic developments of the last 40 years and are of the following nature:

---

[7]The Executive Director of UNEP, Dr. M. Tolba, mentioned such an instance in Tanzania, where 200 tons of pesticides were stored inappropriately and could no longer be used. Since Tanzania did not have the means to dispose of them safely, they had no choice but export them (*The German Tribune* February 26, 1989).

1. The aim of the past political system was to maintain political stability in the context of the steadily increasing military spendings.
2. The efforts to increase the GNP was maintained at the price of neglecting most of the nonproductive areas, including environmental investments:
   a. Rigorous exploitation of natural resources combined with inefficient industries causing high pollution emissions;
   b. Lack of environmental legislation, ineffective administration, and the absence of effective economic stimulations; and
   c. Suppression of information on the real state of the environment and its impact on society (Majak 1990; Environmenal Protection Committee 1990).

Other reasons for environmental damage are a high energy consumption, a desolate economy, the use of old technologies, and a relative high foreign debt (Hilz, Kraus 1990).

With the disintegration of East Germany in late 1989, the then existing "Round Table" of various organizations made an environmental inventory that revealed the following: In East Germany are 121 controlled landfill sites, 4870 controlled dumping grounds, and 7437 uncontrolled and illegal dump sites (IÖW 1990). Furthermore, over the past ten years, more than 5 million tons of waste were imported to East Germany each year, of which West Germany alone imported 650,000 tons of toxic waste and more than 200,000 tons of waste sludge per year. Additional hazardous wastes were imported from Switzerland and the Netherlands (IÖW 1990).

The Round Table, and thereafter the new government, made clear that these waste import schemes will be canceled according to existing laws and provisions from binding contracts. With the unification of the two German states, the majority of this waste trade will end anyhow, as by definition it will only be a movement of waste within national borders. Given the application of West German law in the former East Germany and the grave capacity shortage of hazardous waste landfill sites in West Germany, it can be concluded that no foreign waste will be imported in the united Germany.

The political changes in Eastern Europe allow for an investigation of waste imports into these countries and, with recent information, brief country reports on Poland, Czechoslovakia, Hungary, the USSR, and Romania could be researched.

Poland

Polish television carried in March 1989 a programme on the problem of waste and the public's concern over the import of waste, including radioactive wastes. The Minister of Environmental Protection and Natural Resources noted that there had been attempts to export wastes to Poland from other countries, and that several attempts to illegally import wastes by private persons and enterprises had also been discovered (BBC 1990f). According to the programme, the government of Poland was firmly opposed to this, and the President of the State Atomic Agency said that every import of radioactive substances and waste required an agreement that, so far, has not been granted despite persistent attempts by foreign companies to obtain such agreements (BBC 1990f).

Before the reformation government took power in 1988, the stated policy of Poland was that "banning the export of hazardous wastes completely at this stage would be difficult. However, disposal should be as close as possible to the place of generation and import and exports should only be allowed under very exceptional conditions" (UNEP 1988, 4). The Solidarity majority government enacted an import ban on hazardous wastes that has been enforced since July 1989. The ban, which presents an article in the law on the protection of the environment, contains enforcement action and imprisonment for illegal waste importers. The government also set up a task force on waste trade in order to better identify, track, and prevent waste imports. In early 1990, the state inspectorate of the environment of Poland (PIOS) stated "the new law impotent as the customs officials lack experience and resources to monitor shipment, and that penalties for illegal traffic are too low" (Greenpeace 1990, 40). Poland is thus experiencing the same defenselessness in this regard as many developing countries. These result from vague and different definitions of hazardous wastes, inadequate means of enforcement, and other weaknesses.

In October 1990, Greenpeace International launched an information campaign to inform European legislators about hazardous waste exports from Western Europe to Eastern Europe. In its report *Poland: The Waste Invasion,* Greenpeace revealed 64 known waste trade schemes form 13 countries and 72 firms, amounting to a total of 22 million tons of waste offered to Poland from the end of 1988 to 1990 only. Most of this information was provided by the Inspectorate of the Ministry of Environment Poland (Greenpeace 1990). According to the documentation, of that total, over 46,000 tons of toxic wastes has actually crossed the border, mainly from countries like West Germany, Austria, and Sweden (see Table 5-3). The documentation shows that, since the opening of Eastern Europe to free-market economies, aid packages, and joint ventures, "the dumping of hazardous wastes is, unfortunately, alive and prospering" (Greenpeace 1990, 1).

One proposed waste scheme involves 17,000,000 million tons of toxic sludge and infectious and radioactive wastes. This scheme, which increases the total amount substantially, was offered by the Raremin company of San Jose, California, United States. Mr. Swiatek, an environmental inspector of the Polish authorities said that the proposal had been rejected (Przeglad Tygodniowy 1990; Greenpeace 1990). All in all, the documentation shows that waste exports to Eastern Europe follow the same structural patterns as waste schemes to Africa. Wastes are 'mislabelled as recyclable materials' and involved illegal activities.

TABLE 5-3.   Inventory of Waste Exports to Poland.

Type of Waste	Schemes	Quantified Schemes	Quantity Tons Offered	Quantity Tons Delivered
Miscellaneous, unspecified	17	10	17,275,960	1,920
Waste solvents, paints, liquid chemicals	10	7	5,823	5,207
Metals wastes, electronic scrap	22	13	44,978	12,074
Sewage sludge, industrial and household	4	2	6,200	800
Dredged spoils	4	2	4,350,000	0
Incineration residues, ash, slag, filter cakes	4	2	63,000	27,000
Facility proposals to accept imported wastes	3	0	0	0

Source: Greenpeace. 1990. *Poland: The Waste Invasion, A Greenpeace Dossier.* Amsterdam: Greenpeace International.

In sum, the actually occurring waste schemes involved 7,000 tons of incineration residues from West Berlin delivered to the Polish cement industry and 9,000 tons of PCB-contaminated waste from the Thyssen steelworks. In addition, 15,000 drums of liquid chemical wastes, paint sludge, solvents, and cleaning agents, amounting to 5,207 tons are known to be in various locations in Poland. 50,000 used batteries from separated domestic waste collection in the West "were put on the market in Poland as being 'new' batteries and dried paint residues were sold for 'reprocessing' " (Greenpeace 1990, 3). Even more critical is the trial to sell complete plants for hazardous waste incineration and recycling, sometimes offered for free on condition that the plants accept waste imports and that the residues remain in Poland.

Czechoslovakia

One of the latest Czechoslavakian governmental reports presented to the National Council in early 1990 extensively discussed environmental issues and portrayed some grave environmental problems (BBC 1990e). The summary of the report points out that one of the most dangerous problems is presented by pollution caused by sources that are difficult to monitor, such as waste disposal sites. The report states:

Our country is plagued by tens of thousands of illegal waste dumping sites, many of which can become dangerous sources of pollution of the atmosphere, soil, water, flora and fauna. This ecological crisis already affects more or less our entire republic.…[As a result,] soil erosion threatens approximately 55% of agricultural land, 60% of all soils are acid, the chemically and biologically affected soil produces food detrimental to health, animal species becoming getting [*sic*] extinct, serious genetic changes are appearing among farm animals and also among humans. Forests are damaged to the extent that soon they will be endangered in the whole of our territory, and in the northern third of the country they may disappear altogether (BBC 1990e).

A political assessment of the state of the environment could not be more clear. Although numerous differences exist between the various countries in Eastern Europe, this assessment may indicate to what extent the environments of other eastern countries are polluted and destroyed. A more or less reliable indicator of the state of the environment in the various eastern countries is the average life expectancy at birth. If 1982 and 1984 are compared, it continued to raise in Romania, remained constant in the USSR, and fell in Poland, Hungary, Bulgaria, East Germany, and Czechoslovakia (see Figure 5-1) (European Parliament 1989).

Hungary

Hungary produces 1,800,000 tons of toxic waste per year that requires special handling. Two-thirds of this can be recycled, treated, or made harmless; the remaining third is is kept in temporary storage facilities. Hungary, according to

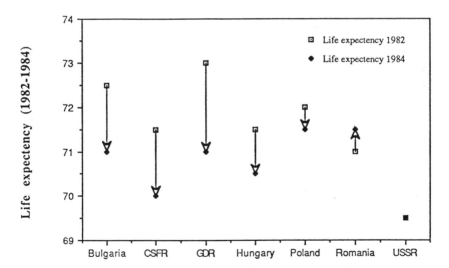

FIGURE 5-1.   Change of Life Expectancy from 1982–1984.   Source: European Parliament. 1989. *Background Note on the State of the Environment in the Countries of Eastern Europe,* Directorate-General for Research. PE 137.119. Luxembourg: European Parliament. Based on data from Statistisches Bundesamt. 1987. *Allgemeine Statistik des Auslandes.* Wiesbaden, Stuttgart: Statistisches Bundesamt.

Hungarian broadcasts, does not have substantial waste exports or imports. A government decree of 1987 prohibits the import of wastes, but waste imports are possible if the wastes are to be used for recycling. The report further points out that Hungary is the transit route for 60 to 80 wagonloads of toxic waste per month (BBC 1989d). Assuming that on average 70 wagons are on transit per month, on a yearly basis this waste scheme would result in 53,760 m3/year. The report does not specify where the waste comes from and where it is shipped. In order to check the transit shipments and/or ban their transport, Hungary has signed the Basel Convention.

USSR

The Soviet news agency TASS reported at the end of 1989 that a regional cost accounting environmental protection system has been introduced in Sumy in the Ukraine (Petrunya, Voronenko 1989). With the restructuring of environmental conservancy in Ukrania, the republic plans to set up special funds to be formed from payments for the damage inflicted to nature as a result of industrial activities, as well as donations. An economic analysis helps define the rates of industrial enterprises' payments for the contamination of air, water, and irrational land use. Starting from 1991, a new economic mechanism of environmental utilization in the USSR will be the basis for wildlife management funds into which enterprises will pay for all polluted discharges into the environment, waste dumping, and hazardous use of natural resources (Petrunya, Voronenko 1989).

Romania

Romania has been the target of waste import schemes involving hazardous wastes from West Germany, Italy, and Austria. Generally, the wastes were transported on the Danube to Romania and finally disposed or dumped in the Black Sea (European Parliament 1990). In another project, a Liechtenstein based company signed an illegal contract with Chimica Bucharest to import petrochemical waste into Romania, Bucharest radio and Agerpress reported. Chaired by former President Nicolae Ceausescu, the Romanian Communist Party (RCP) Executive Committee decided to discipline the responsible officers for "serious violations of the law" connected with this waste import scheme (BBC 1988a). The General Prosecutor's Office reported on an illegal contract with a Liechtenstein company that "deposited large amounts of industrial chemical waste and petrochemical residue in the free port of Sulina," (BBC 1988b) dating back as far as 1986. In 1987, the foreign trade company Chimica Bucharest signed a contract with the same Liechtenstein company to use these products in Romania. Although there are no further details available, the products described as petrochemical products could have been contaminated oil intended for electricity generation in Romania as that country encounters stiff energy shortages. A similar waste export scheme, involving PCB-contaminated oil, was illegally exported from the West Germany to Turkey. As a result of the Executive Political Committee's meeting, the head of the Bucharest port administration and the director of the Chimica Bucharest company were dismissed and two members of the RCP Executive Committee were removed. The meeting further issued warnings to several ministers and to Premier Constantin Dascalescu.

Romania's law prohibits the import and storage of any substance that might endanger the population's safety and health from outside into the country's territory. As a result, the foreign company involved in the above import scheme would be sued according to national and international laws (BBC 1988b). A month later, in July 1988, the trial of those involved in these waste imports began in the Bucharest municipal court. The court revealed that more than 4000 tons of petrochemical products, which upon an investigation were found to be very dangerous to the health of the population and to the environment, entered the country without approval of the National Council for Environmental Protection. The court sentenced Hugo Weinstein, the representative of the Liechtenstein company, to 18 years of prison. Each the 11 Romanian counterparts were sentenced to imprisonment ranging from 11 to 14 years (BBC 1988c).

In sum, Eastern European countries have been realizing the dangers of hazardous waste imports and are establishing either rules for the control of such imports or to prohibit such imports altogether. To what extent these countries can successfully prevent illegal imports remains doubtful.

EXPORT PROCESSING ZONES
AND TRANSNATIONAL CORPORATIONS

Transnational Corporations

Transnational corporations (TNCs) play a crucial role in the development process of LLDCs and NICs and can substantially contribute to the reduction in global environmental risks—both are preconditions for long-term sustainable development. The potential impact of the 2,000 to 3,000 large TNCs' activities can be seen in that they affect at least one-quarter of the world's assets, 80 percent of the world's land cultivated for export-oriented crops, and a major share of the world's technological innovations (UN Commission on Transnational Corporations 1989). In the past decade, 60 percent of industrial investment in LLDCs and NICs originated from outside, much of it coming from transnational enterprises (UNEP 1987). With respect to environmental protection, TNCs can directly and indirectly reduce hazards and improve environmental protection by adopting environmental policy directives and management techniques, improving methods of EIA for high-risk technologies they use or license, disseminating information on environmental standards, and revising procedures for handling toxic wastes (UN Commission on Transnational Corporations 1989).

In the past, differences in environmental standards between industrialized and LLDCs and NICs do not seem to have been a major factor in location decisions. On the one hand, primary considerations for TNCs have included resource characteristics, size of internal market, labor, energy, and transport costs. On the other hand, for LLDCs and NICs, environmental protection has not been a major concern in negotiations, with TNCs leaving them uncontrolled in their activities and environmental performance. The primary issues for LLDCs and NICs in seeking cooperation with and residence of TNCs were foreign exchange earnings, taxation, and employment opportunities for the local workforce (UNEP 1987).

Recent studies by the OECD and UNCTAD indicate, however, a trend of foreign firms taking advantage of "lower costs for meeting pollution abatement and environmental protection standards" (UNEP 1987, 168). As environmental regulations and standards become more stringent in industrialized countries and thereby require high investments in environmental protection, locations in LLDCs and NICs become increasingly attractive. This has proved particularly applicable in heavily-polluting industries, such as nonferrous metals, asbestos, and others (UNEP 1987). The increasing export of hazardous wastes to LLDCs and NICs has been an activity coinciding with the imposition of more stringent laws and tighter control in industrialized countries. The export of entire industries to countries with little or no environmental regulations are thought by some to be a result of this trend (Castleman 1979). Others doubt that extensive "capital flight" occurs (Ashford and Ayers 1985; Leonhard 1984).

TNCs are typical generators of hazardous wastes in LLDCs and NICs. Their activities are subject to local national legislation, including environmental regulation. As shown above, since most LLDCs and NICs have not sufficiently elaborated environmental laws, TNCs may be in compliance with existing laws but may not operate according to higher standards prevailing in more regulated countries. Furthermore, institutional limitations and lack of trained personnel often weaken monitoring and enforcement of the limited regulations that do exist.

Thus, TNCs represent both a threat and an opportunity for managing hazardous wastes. TNCs represent a chance to improve environmental protection through a variety of their activities, which are instrumental to LLDCs and NICs. However, they also represent a threat in that their environmental performance in LLDCs and NICs is not only lagging behind their performance in industrialized countries, but also in that TNCs may have a financial incentive to capitalize on less stringent environmental laws.

Export Processing Zones

Export Processing Zones (EPZs) are another link between industrialized and LLDCs and NICs. Often, they have assisted and facilitated the development process of LLDCs and NICs, their general merits depending on a variety of factors. In general, EPZs represent locations with special status and economic conditions drawing foreign interests and investments. The incentives for firms to locate in an EPZ are similar to the ones for TNCs mentioned above. In addition, there might be special conditions applying in EPZs, such as limited labor regulations or even the right to locate outside the EPZ while taking advantage of EPZ regulations and benefits. Presently, there are some 176 EPZs in operation worldwide, with another 54 in the planning stage, operating in about 40 developing or newly industrializing countries (UNIDO 1989). While there was less interest in EPZs during the period from 1978 to 1983, EPZs appear to have become more attractive as environmental costs increased in the early 1980s in many industrialized countries.

EPZs present special challenges with regard to environmental protection, since it appears that there are many environmental problems in and around EPZs. Given the existing environmental laws and instruments in LLDCs and NICs, EPZs at best, have to comply with laws of the host countries. These weak environmental laws, described in the previous section, have been promulgated primarily to control national industries and thus may not be adequate to control the waste discharge and pollution from high-technology industries. Given that the application of national laws in EPZs is often limited, this could mean that hazardous wastes can legally be dumped in water systems or coastal waters. Furthermore, controlling and monitoring the industries in EPZs may likewise be limited, making it hard, or even impossible, to detect such dumping. Improper disposal of hazardous wastes in EPZs is probably currently taking place and adverse environmental effects from EPZs can thus be expected. To add only one more complication, technically under- or ill-equipped authorities, as well as a lack of updated information about the technologies used in the EPZs, make enforcement very difficult and present a burden to LLDCs and NICs that they can hardly manage.

Although there is no up-to-date evidence that EPZs have been targeted solely for hazardous waste disposal, they could be an attractive location for HWMFs. One of the largest export schemes, the proposal to export hazardous wastes to Namibia, does not specify an EPZ. However, the contractual conditions and infrastructural settings are comparable.[8] The result of this project would be a large-scale HWMF in Africa, managing hazardous wastes from several industrialized countries.

---

[8]See draft contract: 1988/1989. *Concession of importation and storage licenses for waste material on a 50,000 square kilometer site located along the Skeleton Coast in Namibia.* Luanda, Namibia (copy in hand of author).

References

Ashford N. A. and C. Ayers. 1985. Policy issues for consideration in transferring technologies to developing countries. *Ecology Law Review* 12(4):871–906.

Batstone R. et al. 1989. *The Safe Disposal of Hazardous Wastes: The Special Needs and Problems of Developing Countries.* Vol. I. ed. R. Balstone et al. Washington, D.C.: World Bank.

BBC Monitoring. 1988a. Romanian officials sacked over illegal waste dumping. *Summary of World Broadcasts.* Part 2 Eastern Europe.

BBC Monitoring. 1988b. Officials disciplined over illegal waste dumping by foreign company. *Summary of World Broadcasts.* Part 2 Eastern Europe.

BBC Monitoring. 1988c. Romania: sentences in illegal waste dumping case. *Summary of World Broadcasts.* Part 2 Eastern Europe.

BBC Monitoring. 1989d. Hungary joins international toxic waste convention. *Summary of World Broadcasts.* Part 2 Eastern Europe.

BBC Monitoring. 1990e. Czech government report presented by Premier Pithart to National Council. *Summary of World Broadcasts.* Part 2 Eastern Europe.

BBC Monitoring. 1990f. Poland not 'the dustbin of Europe'? *Summary of World Broadcasts.* Part 2 Eastern Europe.

Castleman B.I. 1979. The export of hazardous factories to developing nations. *International Journal of Health Services* 9(4):569–601.

Conteh, J.S. 1989. Personal communication. Chief of Section. ESCAS Dept. of Organization of African Unity.

Du Vivier. 1988. *Les Vaisseaux du Poison: La Route des Déchéts Toxiques,* Paris: Sang de la Terre.

Environmental Protection Committee. 1990. State ecological program—are they enough? *Environmental Policy Review.* Jerusalem.

European Parliament. 1989. *Background Note on the State of the Environment in the Countries of Eastern Europe,* Directorate-General for Research. PE 137.119. Luxembourg: European Parliament. Based on data from Statistisches Bundesamt. 1987. *Allgemeine Statistik des Auslandes.* Wiesbaden, Stuttgart: Statistisches Bundesamt.

European Parliament. 1990. *Situation in Romania.* Notice to Members of the European Parliament, Delegation for Relations with the Countries of South-East Europe, Director-ate-General for Committees and Delegations, European Parliament. PE 140.109. Luxembourg: European Parliament.

Frey, A. E. 1989. International transport of hazardous wastes. *Environmental Science and Technology* 33(5): 509.

George, A. 1988. The dumping grounds. *South,* p. 38.

Greenpeace. 1990. *Poland: The Waste Invasion, A Greenpeace Dossier.* Amsterdam: Greenpeace International.

Hamza, A. 1983. Management of industrial hazardous wastes in egypt. *Industry and Environment.* Special Issue: Industrial Hazardous Waste Management (4): 28–32.

Hannequart, J.P. 1985. The responsibilities of the competent authorities in regard to trans-frontier movements of hazardous wastes. *In Transfrontier Movements of Hazardous Wastes,* pp. 17–38. Paris: OECD.

Hansen, D. March 20, 1989. Hazardous Wastes: Curbs on shipments overseas urged. *Chemical and Engineering News*, p. 6.

Hilz, C., and H.H. Kraus. 1990. Background note on the state of the environment in the countries of Eastern Europe. *Working Paper*. Luxembourg: European Parliament.

Institut für Ökologische Wirtschaftsforschung. 1990. Ökologischer Umbau in der DDR. *Schriftenreihe des IÖW* 36/90. Berlin: ÖIW.

Laurence, D., and B. Whynne. 1989. Transporting waste in the European Community: a free market? *Environment*. 31(6).

Leonard, H.J. 1984. *Are Environmental Regulations Driving U.S. Industry Overseas?* Washington, D.C.: The Conservation Foundation.

Mahajan, S.P. 1989. Hazardous waste management in india: policy issues and problems. *In Hazardous Waste Management,*, ed. Maltezou, Biswas, and Sutter, pp. 286–293. London: Tycooly.

Macak. 1990. The state of the environment in Czechoslovakia. *STOA Working Paper*. Luxembourg: European Parliament.

Maltezou, Biswas, and Sutter. ed. 1989. Editor's Overview of the Expert Workshop. In *Hazardous Waste Management*, United Nations Industrial Development Organization & International Association for Clean Technology. London: Tycooly.

Maurits la Riviere, J.W. 1989. Threats to the World's Water. *Scientific American—Managing Planet Earth*. Special Issue.

Mgbokwere. T.A. 1988. *Statement at the Board of Governors' Meeting on Agenda Item 3—Nuclear Safety Review For 1987*, before the International Atomic Energy Agency. Vienna.

New York Times. March 23, 1989. UN conference supports curbs on exports of hazardous wastes.

Organization for Economic Cooperation and Development. 1985. *Transfrontier Movements of Hazardous Wastes*. Paris: OECD.

Organization of African Unity. 1989. *Plenary Draft Report of the Rapporteur*. Council of Ministers' Fiftieth Ordinary Session. CM/PLEN/DRAFT/RAPT.RPT(L). Addis Ababa, Ethiopia.

O'Sullivan, D. October 3, 1988. Third world has most chemical contamination. *Chemical and Engineering News*, p. 8–9.

Pedersen, J. 1989. *Public perception of risk associated with the siting of hazardous waste treatment facilities*. Dublin: European Foundation for the Improvement of Living and Working Conditions.

Postel, S. 1988. Controlling toxic chemicals. *State of the World 1988*. Worldwatch Institute. New York: Norton & Company.

Przeglad, Tygodniowy. February 11, 1990.

Taipeh Steering Committee. 1989. *Taiwan 2000: Balancing Economic Growth and Environmental Protection*. The Executive Report, Taipeh.

The German Tribune. February 26, 1989. Efforts to halt transport of toxic waste to third world countries. *The German Tribune* 1360, p. 12.

United Nations. 1971. *Development and Environment*. Report of Expert Panel convened by the Secretary General of the United Nations Conference on the Human Environment. Founex, Switzerland.

UN Commission on Transnational Corporations. 1989. Criteria for Sustainable Develop-
    ment Management of Transnational Corporations. *Ongoing and Future Research:
    Transnational Corporations and Issues Relating to the Environment. Report of the
    Secretary General.* 15th Session. E/C.10/1989/1. New York: UN.
United Nations Environment Programme. 1987. *The State of the Environment.* London:
    Butterworths.
United Nations Environment Programme. 1988. *Analysis of Government Responses to the
    Executive Director's Notes on Hazardous Waste Convention,* UNEP/WG.189/Inf., Nai-
    robi: UNEP.
United Nations Industrial Development Organization. 1989. *Seminar on Promotion and
    Development of Export Processing Zones.* PPD.116 (SPEC.). Vienna: UNIDO.
Vorenko, N., and A. Petrunya. 1989. New environmental protection measures. Moscow:
    TASS-Telegraphic Agency of the Soviet Union.
Wolff, A. 1972. The art of progress: development and the environment. *Man's Home: The
    UN Conference on the Human Environment.* United Nations: New York .
World Commission on Environment and Development. 1987. *Our Common Future.* New
    York: Oxford University Press.
Yakowitz, H. 1989. *Monitoring and Control of Transfrontier Movements of Hazardous
    Wastes: An International Overview.* Working Paper. OECD. W/0587M. Paris: OECD.

6

Unilateral Initiatives

The sovereignty of nation-states allows each country to adopt measures to completely ban, restrict, or explicitly approve of the import and/or export of hazardous wastes. Currently, there is no international law prohibiting transboundary movements of hazardous wastes, and national policies regarding waste exports and imports are regulated only according to existing national laws, as discussed in the previous chapter. These laws currently represent the rules upon which the practice of transfrontier waste movements is based. They also provide the framework for bilateral agreements. Transfrontier waste movements generally affect jurisdictions of three countries: the countries of export, transit, and import. According its primary status as an exporting, importing, or transit country, the country may enact specific legislation to control and monitor hazardous waste movements.

EXPORTING COUNTRIES

Potential exporting countries have taken measures to control and monitor transboundary waste movements. These range from completely prohibiting export to LLDCs and NICs, as proposed by the Netherlands, to freely trading wastes among countries (Islam, Smit 1988). Between these extremes are a variety of countries that have discussed the introduction of certain restrictions in their parliaments.

The effectiveness of any restriction depends upon a system to control and monitor transboundary waste movements. Typical measures are a tracking system to monitor the waste stream, using a mandatory manifest or trip-ticket, and regulations to control the generation, storage, and transport of hazardous wastes.

One measure governments have used to restrict hazardous waste exports is to enact laws requiring a system of Prior Informed Consent (PIC).[1] Principally, such a system requires the exporter to notify the importing country of a proposed waste scheme export prior to its implementation. Thereafter, the country of import has the opportunity to consent or to object to the proposal within a given time period. The PIC procedure is designed to ensure that no wastes are being shipped to countries whose governments are not aware of the project. Assuming the country of import raises no objections, this procedure is considered a safeguard, in that it implies that exported hazardous wastes are disposed of in an environmentally sound manner. The assumption is that governments of importing countries, aware of import schemes, would ensure safe disposal. If they could not ensure safe disposal, it was further assumed they would have objected to the wastes' import. These assumptions prove to be wrong. As many hazardous waste export schemes have shown, importing governments often do not have the necessary information about the content, toxicity, and so on, of shipments, nor are they able to make an informed decision. Lack of expertise and trained personnel may represent additional constraints.

Another measure to restrict hazardous waste exports unilaterally is to require disposal of the wastes in a manner technically equivalent to methods and standards applied in industrialized countries or to require "environmentally sound" disposal, whatever that may mean. Some international proposals have called for such an approach on a regional and international level.[2] The difficulties of these proposals lie primarily in their implementation. It would be very difficult to control and monitor the disposal of hazardous wastes abroad, possibly on a different continent. For example, how could standards and regulations for the United States be implemented and monitored in the Caribbean or Africa? And how could an importing government cope with applying different waste regulations stemming from waste imports from different countries? LLDCs and NICs often have lower standards and less stringent regulations than industrial countries. The import of hazardous wastes from different countries with different standards would not only cause confusion, but would make effective control and monitoring very difficult.

---

[1]The system of Prior Informed Consent (PIC) was adopted by the Governing Council of the UNEP in May 1989, and by the Food and Agricultural Organization (FAO) in November 1989. See *London Guidelines for the Exchange of Information on Chemicals in International Trade*, amended 1989. Nairobi. Dec.15-30. 25.51989. It applies to the export of industrial chemicals or pesticides that are banned or severely restricted in the country of its manufacture. PIC makes the export of such substances contingent upon the importing country's written consent. Most developed countries and major producers of chemicals and pesticides have "vigourously opposed" a PIC system over the past 10 years. The International Register of Potentially Toxic Chemicals (IRPTC) is, according to the treaty, responsible for disseminating notifications on the substances that have been banned or severely restricted in their country of origin. See also Environmental Policy and Law 1989, 40.

[2]See discussion in Chapter 7.

The context of the availability and operation of waste disposal technologies in LLDCs and NICs raises another set of complex issues. For instance, PCB-contaminated sludge requires sophisticated technologies, such as high-temperature incinerators, of which none are currently available in Africa (Traore 1989).

To require "environmentally sound" disposal, as called for by the Basel Convention on the Control of Transboundary Movements of Hazardous Wastes and their Disposal[3] seems even more difficult. Although this term is defined,[4] its implementation, even in industrial countries, still remains a difficult task for large amounts of wastes. Furthermore, it appears doubtful whether "environmentally sound" disposal for certain wastes is currently at all available—even in highly industrialized countries.

The extent to which exporting and primarily industrialized countries can adopt effective unilateral restrictions as described earlier is questionable.[5] As these restrictions affect foreign governments and jurisdictions, their implementability and enforceability is difficult to ensure. Unilateral restrictions may work if they are designed to apply to the country of export only, or if they are part of a bilateral agreement. For instance, an industrialized country could prohibit exports and be capable of implementing such a policy. However, industrialized countries may hesitate or abstain from adopting unilateral export restrictions if it is believed that such restrictions would result in an economic disadvantage for its domestic industries. Of course, this is a serious concern, relating to many environmental issues other than hazardous wastes. The potential economic disadvantage of hazardous waste trade restrictions may weigh larger than potential adverse environmental effects. This represents the typical trade-off between economic benefits and environmental protection, which result is usually determined in the political arena. Unilateral policies are thus inadequate for limiting transboundary hazardous waste movements. Regional or international solutions have to be pursued.

IMPORTING COUNTRIES

Importing countries generally have the same regulatory instruments to control hazardous waste movements as do exporting countries. Industrialized countries

---

[3]The international negotiations under the auspice of the UNEP resulted in an international convention, the Basel Convention on the Control of Transboundary Movements of Hazardous Wastes and their Disposal, adopted in Basel on March 21, 1989. From here on, it is refered to as the *Basel Convention*. See Chapter 7.

[4]Article 2 (Definitions) of the Basel Convention states: "Environmentally sound management of hazardous wastes or other wastes means taking all practical steps to ensure that hazardous wastes or other wastes are managed in a manner which will protect human health and the environment against the adverse effects which may result from such wastes."

[5]For a discussion of the U.S. regulatory policy, see sections below.

that have imported hazardous wastes, such as Canada, Great Britain, France, and West Germany, have enacted sophisticated national laws intended to ensure and control safe disposal with available technologies. This is different in LLDCs and NICs, where the generation of hazardous wastes is limited. Since many LLDCs and NICs presently do not have HWMFs for hazardous wastes, and no waste transfer schemes from developing to industrial countries are known, it is unclear where and how these wastes are managed.

According to the available literature and personal investigation, currently there is no LLDC or NIC that has adopted specific legislation to encourage the import of hazardous wastes, except Mexico and East Germany, which have bilateral agreements with the United States and West Germany, respectively. Although countries of Eastern Europe are not typical developing countries, they represent a special case as they are facing severe environmental problems (Hilz, Kraus 1990). Despite this fact, several of them have imported waste to earn crucial foreign exchange. As several waste schemes to Africa and the Caribbean became public over the past few years, governments in LLDCs and NICs have taken legislative action to prevent continued import into their countries. To date, at least 45 countries, representing the African, Latin-Caribbean, and Asian-Pacific regions, have banned hazardous waste imports and adopted penalties for illegal imports (Greenpeace 1989; *New York Times* 1988). Table 6-1 lists the countries that have banned the import of foreign hazardous wastes as of June 1989.

Countries that have taken unilateral steps to ban waste imports intended to protect their environment from the potential risks from hazardous waste management, no matter whether disposal would be according to state-of-the-art technology or simple dumping. These countries disclose that safe disposal of foreign hazardous wastes could not be guaranteed in their country for whatever reasons. This policy reflects a risk-averse attitude, stemming from the fact that they may

TABLE 6-1.   Countries that Ban Hazardous Waste Imports.

1. Algeria	13. Gambia	25. Liberia	37. Tanzania
2. Bangladesh	14. Ghana	26. Libya	38. Togo
3. Barbados	15. Guatemala	27. Mali	39. Tonga
4. Belize	16. Guinea	28. Niger	40. Trinidad/Tob.
5. Benin	17. Guinea Bissau	29. Nigeria	41. Vanuatu
6. Burundi	18. Guyana	30. Panama	42. Venezuela
7. Bulgaria	19. Haiti	31. Peru	43. W. Samoa
8. Cameroon	20. Indonesia	32. Philippines	44. Zambia
9. Comoros	21. Ivory Coast	33. Saint Lucia	45. Zimbabwe
10. Congo	22. Jamaica	34. Senegal	
11. Dom. Repub.	23. Kenya	35. Sierra Loene	
12. Gabon	24. Lebanon	36. Solomon Isl.	

Source: Grenpeace International 1989, The International Trade in Waste: A Greenpeace Inventory.

not be able to assess whether safe disposal in their country is possible, even if it is. A ban thus represents a safeguard against the uncertainties involved in hazardous waste management and can be an effective instrument if national authorities, particularly customs, are able to control maritime transports. Moreover, as the negotiations have shown, the publicly announced ban also strengthened solidarity among the LLDCs and NICs.

COUNTRIES OF TRANSIT

Since almost all major transboundary movements of hazardous wastes are carried by ships, countries of transit may be affected in case of an accident at sea. Even if a ship does not enter territorial waters and follows a route in international waters, such an accident may result in environmental damage in a country near the route. For this reason, the recent negotiations on this subject have raised the proposition of including countries of transit in the PIC system. This means that countries of transit are informed prior to the shipment and can raise objections. Countries of transit have the sovereign right to take steps to prevent ships from entering their national waters, as has been the case in the past; however, their rights in regard to international waters are much further restricted.

In summary, although exporting countries can restrict or prohibit waste exports, individual countries have limited input in solving the problem of transboundary movements of hazardous wastes because of its global nature. Importing countries can protect themselves from unwanted waste imports if they can enforce a previously adopted ban. Countries of transit basically have the same option to protect themselves in order to protect their national territories. However, spills outside territorial waters may nonetheless actually affect their coastline and fishing industries.

THE U.S. POLICY

U.S. Law and Regulations on Transboundary Waste Movements

The Resource Conservation and Recovery Act (RCRA),[6] which regulates the management of hazardous wastes, has the threefold objective of improving the

---

[6]The RCRA was enacted in 1976 and amended in 1984, 42 U.S.C.A. §§6901-6991i. The essential parts of Subtitle C's provisions for hazardous wastes are: (1) to establish criteria to identify HW, §6921(a); (2) to set standards for generators, §6922, and transporters, §6923, regarding storage, treatment, and disposal methods; (3) to set standards for location, construction, and operation of HWMFs, §6924; and (4) to authorize the EPA to establish policies, regulations, and implementation programs to achieve these provisions, §6928. See Findley and Farber 1988, 163.

practice of waste disposal, establishing regulatory control, and promoting resource conservation (Beer 1984). Although RCRA did not explicitly address the export of hazardous waste, the EPA's regulations on generators and transporters applied also to waste exports and therefore represented some control and restriction (Helfenstein 1988). These export regulations of 1980 were minimal, however, and require (1) yearly notification to the EPA before the initial shipment to each country, with identification of the wastes shipped and its importer, (2) a manifest similar to the one applicable within the United States, and (3) confirmation of the delivery of the waste from the consignee to the waste generator (Federal Register 1986). The export regulations did not require reporting the quantity of wastes exported, the frequency of shipments, or the manner of transportation or disposal abroad, nor did the EPA have the authority to prohibit any exports, even when shipments were previously refused by a foreign country (Helfenstein 1988).

As a result of congressional concern over possible loopholes in the control of U.S. hazardous waste movements and potential foreign policy and environmental problems, the RCRA was amended in 1984 by the Hazardous and Solid Waste Amendments (HSWA)[7], giving the EPA more authority to control hazardous waste exports and to coordinate notification with the State Department (Helfenstein 1988).

Notification and Consent Provisions

The HSWA prohibits any exports of hazardous wastes unless the exporter has notified the importing government via the EPA and the importing country has consented to accept the waste shipment.[8] The notification must be sent to the EPA Office for International Activities (OIA) and must contain (1) essential information about the exporter, (2) a description of the waste, (3) the frequency of export, (4) total intended quantity of waste shipped, (5) all points of entry to and departure from each foreign country through which the waste will pass, (6) means of transportation, (7) the manner in which the hazardous wastes will be treated, stored, or disposed of in the country of import, (8) information about the disposal facility, and (9) the name of any transit country (Congressional Federal Register 1987; Handley 1989, 10173). The exporter's notification to the EPA covers exports of a particular hazardous waste for a 12-month period. Any substantial change in

---

[7]The new provisions became effective November 1986, with the EPA responsible for promulgating regulations necessary to implement the new provisions within one year from the date of enactment. HSWA (1984). Pub. L. No. 98–616. §245.98 Stat. 3221. 3262–64 (codified as amended at 42 U.S.C.A. §§6921–6923. 6938). See also Kelly 1985.

[8]This essentially represents the PIC procedure, as described earlier, 40 C.F.R. §262.53.

data or information of the original notification requires renotification and renewed consent from the importing country.

Within 30 days after receiving notice of a proposed export, the EPA, in conjunction with the State Department, must provide a complete notification to the government of the importing country, including information that the U.S. law prohibits the exports of hazardous wastes without the consent of the receiving country.[9] Furthermore, the receiving country must also be provided with a description of the federal regulations that would apply to the treatment, storage, and disposal of hazardous waste in the United States (Handley 1989). After the EPA receives the consent or objection, it is forwarded to the exporter who may proceed with the export if the country of import has consented. If the shipment is objected to, or if no consent has been received, the EPA cannot grant an export permit. Under the U.S. system, the notification is extended to countries of transit; however, no consent from these countries is required prior to export (Helfenstein 1988).

Manifest and Annual Reporting to the EPA

The EPA has also adopted a manifest system, both within the United States and for transboundary movements of hazardous wastes. Under the manifest system, a transporter may not accept a shipment without a manifest from the waste generator specifying the facility for storage, treatment, or disposal of the wastes. The manifest and consent must accompany the shipment, and a copy must be delivered to customs at the time the shipment embarks. If the exporter knows that the shipment does not conform with the manifest or consent, the exporter must not proceed with the export. In case of violations of export regulations, they are subject to criminal penalties (Helfenstein 1988).

The HSWA also requires each exporter of hazardous waste to file an annual report with the EPA on their activities. The reports summarize the types, quantities, frequency, and ultimate destination of all hazardous waste exported in the previous year. The definition of hazardous wastes as defined in the RCRA[10] includes certain solid wastes, as well as those that exhibit certain harmful characteristics.[11] The regulations further require a certification by the exporter that the information contained in the annual report is true, accurate, and complete.

---

[9]The State Department transmits the notification received from the EPA Office for International Activities to the U.S. embassy in the receiving country. When the government accepts the import, or objects to it, the procedure is reversed and the information transfered back to the EPA-OIA. See also Handley 1989, 10173.

[10]"Hazardous Waste" is defined in RCRA §1004(5). 42. U.S.C. §6903 (5). ELR Stat. RCRA 004 and 40 C.F.R. §261. 1987.

[11]Those harmful characteristics include ignitability, corrosivity, reactivity, or toxicity. See also Helfenstein 1988, 783.

Provisions for Enforcement

"Primary exporters"[12] who initiate hazardous waste exports, as well as transporters and intermediaries that arrange exports, are subject to responsibilities under HSWA export regulations. These responsibilities include timely, accurate, and complete notifications to the EPA, obtaining the written consent of receiving countries, and submitting the annual report. Furthermore, the primary exporter must make specific efforts to verify that the waste was transported to the location specified (Helfenstein 1988).

While the United States has provisions authorizing the government or private citizens to seek injunctions and civil penalties against generators, transporters, owners, and operators of hazardous waste management facilities (HWMFs) who endanger public health or the environment, these provisions are no longer applicable once the hazardous waste leaves the United States[13] Current RCRA and CERCLA provisions are not applicable abroad and do not have any extraterritorial effect (Handley 1989, 10174). For that reason, it does not seem that the "United States government, a United States citizen, or the citizen of a foreign country has any standing to invoke RCRA or CERCLA remedies on the basis of hazardous waste disposal conducted abroad" (Handley 1989, 10174). Accordingly, while the U.S. government or a citizen could file a suit requiring primary exporters to comply with RCRA notification and reporting regulations, they "are powerless to make firms accountable under American law for their hazardous waste disposal activities outside the United States" (Handley 1989, 10174).

Administrative Responsibility

According to the provisions of HSWA, the EPA is the principal governmental authority to promulgate regulations controlling transboundary movements of hazardous waste. The notification and consent procedures are handled in cooperation with the State Department. Moreover, according to the legislative history of RCRA, Congress intended the EPA to collaborate with the U.S. Customs Service "to establish an effective program to monitor and spot check international shipment of hazardous wastes to assure compliance with requirements of the [RCRA] section" (Handley 1989, 10174).

---

[12]This term refers only to those who initially decide to export the HW or to brokers who arrange the export and disposal of HW abroad (C.F.R. pt. 262.51. 1987).

[13]RCRA §7003 authorizes the government and RCRA §7002 authorizes citizens to seek such injunctions. The government may also seek response costs and natural resource damages under §107 of the Comprehensive Environmental Response, Compensation, and Recovery Act (CERCLA) U.S.C.A. §§9601–9675, §9606. See also Handley 1989.

**Recent Legislative Developments
in Controlling HW Exports**

Since 1986, when the revised EPA regulations became effective, and especially since a number of illegal hazardous waste exports focused public attention on the export problem, several legislative initiatives were proposed in the U.S. Congress, but none has been enacted to date.

In 1988, U.S. Senator Kasten (R-Wis.) introduced the "Waste Export Control Act," which would prohibit waste exports, unless disposal facilities meet the standards equivalent to those in the United States (U.S. Congress 1988a). Waste exporters would have to apply for an export permit and be required to provide a "detailed description of the manner in which waste will be transported, treated, stored, and disposed of in the receiving country," as well as provide information demonstrating that exported wastes are managed in a manner "providing environmental protection no less strict than is provided by the requirements...within the United States" (U.S. Congress 1988a). The EPA would be the primary authority to issue permits and also be responsible for the other procedures under the HSWA.

Also in 1988, "The Waste Export Prohibition Act" was introduced by U.S. Representative J. Conyers (D-Mich.), which would ban all exports of solid and hazardous wastes, except if a bilateral agreement *exists* between the United States and the proposed country of import, such as Canada and Mexico ((U.S. Congress 1988b). The law would set penalties similar to the ones already in existence for illegal exports. Current penalties generally apply for violations of EPA regulations concerning hazardous waste exports.

Other bills introduced were the "Waste Minimization and Control Act of 1988" (U.S. Congress 1988c) and the "Waste Act to Save the Environment" (U.S. Congress 1988d), which were introduced by U.S. Senator M. Baucus (D-Mont.) and U.S. Representative J. Porter (R- Ill.), respectively. Both laws would require bilateral agreements similar to Conyers's bill to allow hazardous waste exports. However, the Baucus and Porter bills differ from the Conyers bill in that they would allow the United States to enter into additional bilateral agreements beyond those already in place (Handley 1989).

The latest bill, the "Waste Export Control Act" (U.S. Congress 1989a), was introduced in the spring of 1989, with the purpose of limiting the export of hazardous wastes to exceptional situations. It would prohibit waste exports, unless an international agreement is in effect to which the United States and the receiving country are parties. In the case of export, the management of hazardous wastes would have to meet the same standards as the Kasten bill, and be disposed of under regulations no less strict than in the United States. Under this law, an exporter could be refused a permit to export if "reasonable efforts to eliminate or minimize waste generation prior to export" were not undertaken (U.S. Congress 1989a, 15). Furthermore, the bill would establish liability for any response (i.e., cleanup) costs or natural resource damages, as referred to in Section 107 of the CERCLA of 1986, and give a foreign government the right to bring action against an exporter in the United States (U.S. Congress 1989a). Besides these bills, there have been several congressional hearings on hazardous waste exports, including the potential adverse effects on the environment and human health.[14]

Evaluation of U.S. EPA Regulations

Four years of experience with the 1984 RCRA amendments, under which the EPA promulgated revised export regulations that became effective in November 1986 have revealed a set of weaknesses and loopholes. The major problems are the following:

1. Above all, the EPA's authority is too limited. Currently, the EPA has no authority to stop (and thus cannot stop) an export if there was proper notification and if the importing government gives its consent. Even if the EPA believes that the particular export is dangerous and may cause harm, it is bound to allow the export (*International Environment Reporter* 1988).
2. The notification and consent procedures do not cover wastes that are not considered "hazardous" in the United States, but that are legally defined as "hazardous" abroad.[15] As a result, "the tendency to export solid waste classified as non-hazardous under RCRA is increasing and beginning to pose environmental, health, and diplomatic problems..." (U.S. Congress 1989, 1), as noted by Subcommittee Chairman Mike Synar (D-Okla.) at a congressional hearing.
3. Hazardous waste export requirements were not adequately enforced. The EPA Inspector-General "found instances where hundreds of tons of hazardous waste were exported without notifications".[16] Furthermore, "enormous quantities of hazardous waste were exported without exporters filing the required annual reports" (U.S. Congress 1989b, 21). By 1986, less than 20 annual reports were filed with the EPA, despite the receipt of "hundreds" of notifica-

---

[14]At least two hearings were held by the Environment, Energy, and Natural Resources Subcommittee of the Committee on Government Operations, U.S. House of Representatives, on July 14, 1988, and in May of 1989.

[15]The "EPA interpreted congressional intent in enacting HSWA as not requiring the notification and consent provisions to apply to wastes not regulated within the U.S." Thus, wastes proposed for export and defined as hazardous abroad but not in the United States were not controlled. The EPA Inspector-General testified before the U.S. Congress in 1988 that the current notification and consent procedure should be broadened, to cover municipal waste, including those not legally defined as "hazardous" in the United States See U.S. Congress 1989, 12–15; Handley 1989, 10177–180.

[16]The EPA office of the Inspector-General compared 80 annual reports from 1985–1986, which

tions each year.[17] As a result, the EPA "did not know the amount of hazardous wastes actually exported to other countries" (U.S. EPA 1988, 3).

4. Furthermore, the EPA did not have an enforcement strategy and failed to coordinate its efforts with those of customs. Out of 274 manifests that the EPA National Enforcement Investigation Center (NEIC) received as of December 1987, 143 manifests did not specify the port of exit. The EPA Audit report concludes that "consequently, hazardous waste exporters could disregard EPA regulations with little chance of detection" (U.S. EPA 1988, 4). As a result, if exporters did not provide the required notification, the EPA could not identify and prosecute violators. This practice had the consequence that importing countries were denied the right and opportunity to reject the wastes. In addition, "the receiving country's consent, which EPA forwards to the exporter for attachment to the manifest, did not always contain the data that customs needs to ensure the shipment is proper" (U.S. EPA 1988, 4).

5. The EPA's hazardous waste export regulations were unclear or ambiguous and resulted in the misclassification of hazardous wastes as materials for "recycling" and "reclamation."[18] This practice led to "sham recycling" and "illegal" (U.S. EPA 1988, 5) exports. Another problem stems from the fact that exporters claimed that their shipments did not contain hazardous wastes, but contained economic goods or "products" (Castleman, Navarro 1987).

6. The system of Prior Informed Consent (PIC) did not always work. A waste export scheme to the Congo reveals that the country first consented to the export, but withdrew its consent 10 days later. The Congolese government claimed that the information provided was insufficient to make an informed and sound decision, and if they had known all the information, they would have rejected it at first (Traore 1989). The EPA's National Enforcement Investigation Office supports this claim by stating that hazardous waste manifests show "serious problems with proper completion of these documents and classification of wastes" (U.S. EPA 1988, 19). At the congressional hearing, Representative Synar concluded that current application of *prior informed consent* "may not be as informed as it should be" (U.S. Congress 1989b, 283).

7. The current legal requirements are not enforceable abroad and may lead to different standards regarding the disposal of wastes. Furthermore, it appears that there are no insurance requirements covering improper disposal and accidents abroad.

---

[17]The number of notification increased from 12 in 1980 to over 600 in 1988. See U.S. EPA 1988, 16.

[18]120 drums of hazardous wastes containing mercury sludge were exported to South Africa, being described for "recycling." Likewise, over 11,000 tons of lead flue dust, lead furnace slag, and lead press coke were exported to West Germany, being described as "reclamation." See U.S. Congress 1989a, 22-23.

Besides these problems, there are a variety of other issues that affected the U.S. policy's success. The entire EPA program on hazardous waste exports was not adequately funded or staffed. One single person in the EPA was assigned to handle the program. Moreover, the assignment of three agencies—the EPA, the customs service, and the State Department—resulted in a bureaucratic burden, with lack of coordination and final failure of effective control (U.S. Congress 1989b, 283).

References

Beer, L.P. 1984. Waste disposal/management option. *Hazardous Waste Disposal.*

Congressional Federal Register. 1987. 40 C.F.R. §262.53(a)-(b).

Castleman B.I., and V. Navarro. 1987. International mobility of hazardous products, industries, and wastes. *Annual Review of Public Health* 8:1-19.v.

Environmental Policy and Law. 1989. PIC—abreak-through in north-south relations. *Environmental Policy and Law* 19:2.

Federal Register. 1986. Hazardous waste management system: exports of hazardous waste. *51 Federal Register.* 28.664.

Findley, R.W., and D.A. Farber. 1988. *Environmental Law in a Nutshell.* Minnesota: West Publishing Company.

Greenpeace International. 1989. *International Export of U.S.* Waste. Hearing before the Subcommittees on International Economic Policy and Trade, and on Human Rights and International Organizations, House of Representatives, U.S. Congress. Washington, D.C. July 12, 1989.

Handley, F.J. 1989. Hazardous waste exports: a leak in the system of international legal controls. *Environmental Law Reporter* 89(4):10171-182.

Helfenstein, A. 1988. U.S. controls on international disposal of hazardous waste. *The International Lawyer* 22(3):775--790.

Hilz, C., and H.H. Kraus. 1990. *State of the Environment in the Countries of Eastern Europe*, Directorate General for Research, European Parliament, Luxembourg.

International Environment Reporter. August 10, 1988. EPAs program on hazardous waste exports needs improvements, Inspector-General says. *International Environment Reporter.*

Islam S., and P. Smit. 1988. Dirty games in Brussels. *South,* p. 37-41.

Kelly, M.E. 1985. International regulation of transfrontier hazardous waste shipments: a new EEC environmental directive. *Texas International Law Journal.* 21(67):85-128.

New York Times. September 22, 1988. African nations banning toxic wastes. *New York Times*

Traore, A. 1989. Le sud ne doit pas etre la poubelle du nord. Interview with Francois Roelants du Vivier. *Le Courrier.* 113.

U.S. Congress. 1988a. House of Representatives. S. 2598. 134 Cong. Rec. S. 8809-10.

U.S. Congress. 1988b. House of Representatives. 5018. 100th U.S. Congress.

U.S. Congress. 1988c. House of Representatives. S. 2773. Rec. S 12289.

U.S. Congress. 1988d. House of Representatives. 5434, 100th U.S. Congress.

U.S. Congress. 1989a. House of Representatives. 2525. 101st U.S. Congress.
U.S. Congress. 1989b. *International Export of U.S. Waste. Hearing before a Subcommittee of the Government Operations*, House of Representatives, 100th Congress, July 14, 1988. Washington, D.C.: Committee on Government Operations.
U.S. EPA. 1988. EPA's *Program To Control Exports of Hazardous Waste*. Report of Audit E1D37-05-0456-80855. Washington, D.C.:U.S. EPA.

7

International Policies

This chapter will examine *international* or multinational policy propositions to control transfrontier movements of hazardous wastes. International approaches can be divided into three levels: bilateral, regional, and global policies. Each of these levels includes three basic options: a complete ban, limited exports or imports, and unrestricted trade. A combination of these levels and options would basically result in more than ten theoretical options; however, only a limited number of these are practical and feasible. The policies described and analyzed below represent the options that emerged from the discussions and negotiations on all three levels.

BILATERAL AGREEMENTS

Bilateral approaches to solutions for hazardous waste management provide many opportunities for both importing and exporting countries. Whether they are the optimal tool to control international movements of hazardous waste needs to be examined. The following two sections give an analysis of bilateral approaches taken by the United States, which was selected because it is the largest manufacturer of chemical substances and hazardous wastes, as well as the largest exporter of hazardous wastes. Furthermore, along with the Netherlands and the West Germany, the U.S. has sophisticated safeguards to control and monitor hazardous wastes (UNEP 1983).

The United States represents a unique case in that it has two bilateral agreements, one with an industrialized country (Canada) and another with Mexico.

While Mexico has a large potential for future industrial growth, it is also confronted with serious environmental problems resulting from its past development. Having experienced rapid industrial development over the past two decades, the institutional infrastructure that controls and monitors environmental effects and counteracts environmental degradation could hardly keep pace. As a result, the system for environmental protection in Mexico is unlike the one in Canada. The two bilateral agreements of the United States with Canada and Mexico will be examined below. Then, the question of whether countries with differently developed economies are prepared to effectively carry out bilateral agreements to control hazardous waste movements will be addressed.

U.S. Bilateral Agreement with Canada

In 1986, the United States signed a bilateral agreement with Canada on hazardous waste movements. The 14-article agreement (U.S. EPA 1988) not only covers U.S. waste exports, but also covers imports of hazardous waste from Canada and transit shipments of waste through Canada between two U.S. points. The agreement's requirements regarding monitoring of hazardous waste movements are slightly different in several aspects than the HSWA's requirements, described earlier.

The essential provisions of the agreement with Canada are Notification of Intent and PIC procedures (Article 3), allowing the receiving country 30 days from the date it receives the notice to object or consent to the shipment. If no response is received within 30 days, consent is *implied*. The agreement further requires that shipments conform to the regulations of the *receiving* country. This requires, according to articles 3 and 8, that U.S. waste generators must register with Canada and provide the same information as is required from Canadian generators. Article 4 calls for notification of transit shipments and does not include a provision for objection. Another provision (Article 5) calls for cooperative efforts in monitoring and spot-checking in order to ensure compliance with regulations of both countries. The agreement also includes a provision for readmitting exports (Article 6) if the receiving country wants to return the waste. Parties may also require, as a condition of entry, that any transboundary movement of hazardous waste is insured against damage to third parties (Article 9).

The comparison of the EPA regulations for general hazardous waste exports and the special bilateral agreement between the United States and Canada shows significant differences in several aspects. Before some of these will be evaluated in the context of their application, a look at the bilateral agreement between the United States and Mexico should be of interest.

U.S. Bilateral Agreement with Mexico

In 1986, the United States also entered into a bilateral agreement with Mexico.[1] The agreement also allows the import of hazardous waste from Mexico into the United States, as well as transit shipments through the United States and Mexico. The regulations for notification and PIC are slightly modified, and special provisions apply to hazardous waste generated by U.S. companies manufacturing in Mexico.

The agreement requires the exporting country to provide a notification of intent to export hazardous waste to the importing country 45 days in advance of shipment, and the response of consent or objection to the importing country shall be reported in another 45 days (Article 3). In contrast to the Canadian agreement, if a response from the importing country is not received within the prescribed time, consent is *not* implied. Transit shipments require prior notification, but not consent (Article 4). According to Article 2, Sec. 3, each party "shall" cooperate in monitoring and spot-checking and in the exchange of information (Article 12). The agreement also includes a specific provision for raw materials that were brought from the United States into Mexico for manufacturing. If the manufacturing process generates hazardous waste, these "shall...be readmitted" by the country of origin of the raw material (Article 11). The liability provisions call for the country of export to take the following steps, in case of a violation of national laws and other conditions further defined:

1. Return the hazardous waste...to the country of export.
2. Return, as much as practicable, the *status quo ante* of the effected ecosystem.
3. Repair, through compensation, the damages caused to persons, property, or the environment (Article 14).

The same article calls for insurance coverage of waste shipments. The export of hazardous waste to Mexico is limited to wastes that can effectively be recycled. Exporters must also provide a manifest for each shipment and annual reports to the EPA (Rose 1989).

Although this agreement allows the exports of hazardous waste to Mexico, a presidential decree (U.S.) (Rose 1989) prohibits any imports of hazardous wastes to Mexico for disposal and is thus superceding the agreement. The bilateral agreement with Mexico also needs to be interpreted in the context of Mexico's

---

[1]Agreement of Cooperation Between the United States of America and the United Mexico States Regarding the Transboundary Shipments of Hazardous Wastes and Hazardous Substances. See U.S. EPA 1988. Appendix C. In addition to HW, Articles 5 to 9 of this agreement specifically cover "hazardous substances," such as banned and severely restricted pesticides and chemicals.

Maquiladora Program,[2] set up to attract U.S. industries in order to promote industrial development in northern Mexico. Although the Maquiladora program requires that waste residues from processing are returned to the United States for disposal, often, hazardous waste are classified as "recyclable materials [wastes are only allowed to be exported to Mexico for recycling] destined for recycling when the recycling process actually involves disposal" (Rose 1989, 228). The EPA Inspector-General reported at the 1988 congressional hearing of such an instance involving 10,000 gallons of heavy hydrocarbons and other hazardous materials (U.S. Congress 1989, 24).

In sum, the bilateral agreement with Mexico allows hazardous waste exports, but is severely restricted by the presidential decree. As a result, only hazardous wastes that are to be recycled are legally allowed to be exported to Mexico. However, according to the EPA Inspector-General's testimony, hazardous wastes were often classified as "recyclable" materials, although they were actually disposed.[3]

Comparison of the Canadian and Mexican Agreements

In comparing the two bilateral agreements with Canada and Mexico, two main conclusions can be drawn: (1) Combined with the presidential decree to prohibit waste exports to Mexico that are designed for disposal, consent provisions of the bilateral agreement with Mexico are clearly more restrictive than the Canada agreement. While the Canada agreement implies consent, if no response is received within 30 days of notification of the competent authority of the importing country about a waste shipment, consent is *not* implied in the Mexico agreement if no confirmation is received within 45 days after notification. (2) Procedures and regulations regarding control and monitoring of waste movements appear to be of equal strength in both agreements. Basically, the agreements cover the same provisions regarding notification, consent, enforcement, cooperation, and liability. The language in both agreements is similar, and one could not conclude that one agreement is stronger than the other on its face.

In practice, however, it appears that there have been many more problems with the Mexican agreement. Reports covered several cases of "sham recycling" and "illegal" shipments of hazardous wastes to Mexico (Rose 1989, 224–225). Problems with hazardous waste are supported by the fact that on the U.S. side of the U.S.-Mexican border there are numerous serious pollution problems stemming in

---

[2]This program was established in 1965 as a solution to the problems associated with economic underdevelopment in northern Mexico. It is a border industrialization program with the United States, to attract U.S. industries (maquiladoras) and includes (1) duty-free entry of equipment and materials, (2) duty-free and subsequent tax-free reexport, and (3) low minimum wage. See Rose 1989.

[3]See "Evaluation of U.S. EPA Regulations" section in Chapter 6.

part from hazardous waste dumping in Mexico's Maquiladoras. Furthermore, pollution from hazardous wastes may also be a result of Mexico's limited capacity to dispose of hazardous wastes, a reason for uncontrolled dumping.

Currently, about 80 percent of all U.S. hazardous wastes exported go to Canada, and up to 90 percent of the U.S. waste that is exported to Canada has been managed in only two incinerators. The system of PIC appears to have worked effectively in the U.S.-Canada agreement. This is evidenced by the fact that Canada raised 61 objections to intended shipments from 54 companies within a five-month period in 1987 (U.S. Congress 1989). In addition, Canada's "administrative efficiency" and "high level of environmental supervision" (Helfenstein 1988) allows tighter control over disposal, as well as a faster response in case of an emergency or a violation.

Evaluation of Bilateral Agreements

The evaluation of the U.S. EPA regulation identifies several major problems described above. Although some of these problems and weaknesses could be solved in bilateral agreements, still others impede effective control. The following conclusions emerge:

1. Whereas the U.S. EPA regulations do not cover hazardous wastes not defined under RCRA, the bilateral agreements' procedures should conform to the regulations of the receiving country. Thus, all wastes deemed to be hazardous in either one of the countries should be included in the agreement.
2. Notification and consent procedures are not equally successful in all circumstances. They appear to have worked effectively in the bilateral agreement with Canada, although Canada was not notified of some U.S. exports. The number of exports without notification were small (The EPA Inspector-General identified nine exports without notifications between 1985–1986, during which approximately 500 shipments went to Canada). Since the majority of wastes to Canada goes to two facilities only, the authorities of both countries can better track, control, and monitor the waste stream. In addition, Canada has successfully objected to 61 exports to their country, which is an indication that these procedures in the bilateral agreement work. Notification and consent procedure appear to be less successful with wastes sent to Mexico. They could be even less successful for hazardous wastes sent to LLDCs and NICs.[4] As the

---

[4]The U.S. EPA Inspector-General provides a list of countries to which U.S. HWs have been sent in the past; they include Canada, Mexico, England, West Germany, Philippines, South Africa, Taiwan, Australia, Belgium, Brazil, Holland, India, Japan, Korea, Spain, Hong Kong, Dominican Republic, Columbia, Venezuela, and Thailand. See U.S. EPA 1988, 18–21. According to Greenpeace, several other countries in Africa, Latin America, and the Caribbean have received HWs from the United States.

shipment to the Congo shows, the country was not properly notified. A study on the EPA notification programs found in 1985 that "developing countries have trouble obtaining adequate information for decision making; information still does not reach the appropriate people in most instances..." (Halter 1987, 26; Helfenstein 1988, 789). As a result, as in the case of the Congo, consent may be premature and based on inadequate and insufficient information.

3. The bilateral agreements with Canada and Mexico have specific provisions for liability and insurance, and thus allow compensation in case of damage to people or the environment, whereas the current EPA regulations do not include such provisions. It should be noted that both bilateral agreements also allow the import of hazardous wastes into the United States, which might be a reason why liability provisions were included. This is quite distinct from the EPA regulations, where it was assumed that hazardous wastes are almost exclusively exported. If hazardous waste imports were anticipated by the United States, liability and insurance provisions most likely would have been included, as they are standard in U.S. hazardous waste management. The general inclusion of liability provisions in HSWA would result in several additional requirements regarding control and monitoring provisions, and would in practice further complicate hazardous waste exports.[5]

4. Finally, the current U.S. legislation does not allow the EPA to stop any exports if the shipment is "agreed to" by the importing country. Even if the EPA is convinced that the wastes might be dumped or otherwise mismanaged, the agency has no authority to prevent a waste shipment. In contrast, the bilateral agreements have specific provisions allowing the importing country to stop a shipment and thus prevent potential environmental harm.

The comparison of the bilateral agreements with Canada and Mexico reveals that the Canadian practice generally meets the conditions set forth in the agreement. It can be further concluded that imported hazardous wastes to Canada are disposed of according to state-of-the-àrt technology and in accordance with regulations. The agreement with Mexico allows hazardous waste exports to Mexico intended for disposal, although this is forbidden by a presidential decree. Provisions for the reimport of hazardous wastes that were exported to Mexico as raw materials further complicate the regulations and exceptions. Furthermore, hazardous wastes classified as recyclable materials were exported to Mexico and "sham recycled." Given the infrastructural weaknesses, the complexity of regulations and exceptions, and the practice of sham recycling, it can be concluded that hazardous wastes imported to Mexico are not effectively controlled and monitored. Certain other infrastructural differences are also likely to contribute

---

[5]See "An EC Directive on Liability," section in Chapter 7 and Chapter 8, recommendation 6.

to why hazardous waste exports to Mexico are not handled as designed in the agreement. These differences, as well as economic and bureaucratic differences between Canada and Mexico, may partly explain why one agreement works better than the other.

The experience with the Canada and Mexico agreements show that bilateral agreements between highly industrialized countries can work effectively, while agreements between a highly developed country and a country in the development process may not work as well. The bilateral agreements are better equipped to govern transboundary movements of hazardous wastes than are the present EPA regulations. Major improvements in these regulations are necessary in order to meet the conditions of a sound export policy.

GLOBAL AND COMPLETE BAN

A number of countries, many of which have banned the import of hazardous wastes into their territories, have supported a global ban[6] of transboundary movements of hazardous wastes. The member countries of the OAU also supported a global ban during their 1988 annual conference (OAU 1988). The need for such a ban was further endorsed and supported by numerous national, regional, and international environmental Non-Governmental Organizations (NGOs).[7]

A policy of a global and complete ban on hazardous waste exports and imports can be adopted only by an international forum with the institutional capacity and authority to coordinate the supervision of the policy's implementation and enforcement. The setting of such international approaches will be discussed in the next sections in regional and international negotiations. A complete ban of hazardous waste exports would have consequences for both industrialized nations and LLDCs and NICs. The following main arguments for and against a ban examined from the perspective of industrialized countries and developing countries should contribute to a better understanding of the merits and deficiencies of such a policy.

---

[6]A global ban means a complete prohibition of *any* exports and imports of HWs on a global level. Over the past years, developing countries called for a global ban to developing countries, as distinguished from a global and complete ban. In effect, a ban limited to developing countries would not represent a global ban and thus should not be characterized as such.

[7]Notably, these include Greenpeace International, which investigated hundreds of HW export schemes and provided information and data to the ongoing UNEP negotiations, as well as to many developing countries. Greenpeace's investigations and its international network was able to publicize illegal shipments and inform the public. These activities have indirectly helped prevent ongoing shipments from unloading its cargo and has also helped stop other illegal shipments.

Industrialized Countries

Many industrialized countries are currently exporting, and a lesser number of countries are importing, hazardous wastes. In some industrialized countries, the capacity to manage and dispose of hazardous wastes is very limited. These countries are currently forced to export their wastes if they do not want to dispose of them in an environmentally dangerous or illegal manner. The Netherlands and West Germany are particularly affected countries. The Netherlands needs to export hazardous wastes because of their special hydro-geological conditions. West Germany needs this option because of its limited current waste disposal capacity. As a result, if exports as a disposal option did not exist, a "national state of [waste] emergency" would be a matter of reality (Frankfurter Allgemeine 1989). Thus, a gradual ban or one put into effect far enough into the future could stimulate effective waste minimization and source reduction approaches.

As shown in Chapter 5, the practice of hazardous waste exports and imports among industrialized countries has resulted in economic benefits. Assuming that the decisions of European countries to trade hazardous wastes were guided by the principle of disposing the hazardous wastes in as environmentally sound a manner as possible, it could be argued that the exports of hazardous waste to other European countries may also have resulted in better environmental protection. As a result of these benefits, which are also a result of economies of scale, the trade in hazardous wastes among member countries of the EEC, as well as between the United States and Canada has been increasing over the past decade. Industrialized countries want to continue this trade and are against a complete ban, which could cause economic inefficiencies and which may even have adverse environmental consequences in the short run.

There are other important consequences to a global ban on hazardous waste exports. Currently, many industrialized countries are facing challenging problems with increased waste generation, limited disposal capacity, and adverse public attitudes towards new facility sitings. In combination with increasing environmental pressure, these countries have, to different extents, engaged in waste reduction, waste recycling, and reuse of waste. Although many countries have adopted policies to encourage such activities and are making some progress, many economic, political, and institutional obstacles still impede waste minimization and source reduction efforts. Mounting political and public pressure on industries to improve their environmental performance most likely contributed to the industries' decision to opt for hazardous waste exports to LLDCs and NICs, besides being a lucrative business as well.

Therefore, a total ban on hazardous waste exports has to be viewed in light of how a complete ban would affect waste minimization policies. For some countries, such a ban would have significant consequences. Because of the shortage of

HWMFs, some countries would be forced to drastically reduce their waste generation while recycling and reusing as much waste as possible from the rest. Environmentalists argue that a complete ban on exports would thus serve a good cause. While contributing to a better use of resources, environmental protection would increase because less hazardous wastes are generated. In sum, this whole process would contribute to a more sustainable society.

However, many industrialized countries would find themselves with large volumes of hazardous wastes piling up, without any practicable solution in the short term. A possible consequence could be that sea incineration and especially sea dumping would again be heavily utilized until other long-term solutions might become available. This scenario could result in significant environmental damage. On the other hand, given the self-sufficiency principle, adopted by the EC, such a ban would represent a serious signal to industry that all hazardous wastes would have to be handled within the country of generation. In the short term, this may cause a number of problems regarding the current management; on the long term, however, it would lead to a responsible policy of self-sufficiency.

Developing Countries

A global ban would provide LLDCs and NICs some protection against hazardous waste shipments. This security could provide important psychological reassurance of their sovereign rights and their emancipation from colonial and post-colonial paternalism. All future shipments would be illegal. Whether such a ban could indeed be implemented and controlled, and thus provide actual protection, remains to be examined.

A global and complete ban would certainly recognize the vulnerability of LLDCs and NICs and give consideration to their current limitations to effectively manage hazardous wastes. It needs to be recalled that LLDCs and NICs are confronted with increasing problems with hazardous wastes, and it would be difficult for them to assure the additional responsibility of managing foreign wastes. Given the constraints to proper waste management in LLDCs and NICs, a global and complete ban would contribute little to environmentally sound management of their own wastes. In the worst case, such a ban would even prevent a shipment of highly toxic waste from a developing to an industrialized country, as might be necessary if a developing country has to dispose of a highly toxic waste for which it has no environmentally sound disposal method, and for which a proper disposal method exists in another country.

A complete ban will certainly have other effects with regard to international cooperation, information, and technology exchange on hazardous waste management. The current process of information and data exchange included in many international agreements would no longer be required, and this might adversely effect LLDCs and NICs in their technical ability to cope with their own waste problems.

Evaluation

A policy to completely ban transboundary movements of hazardous wastes on a global level could have serious effects on both industrialized nations and LLDCs and NICs. The consequences of such a ban would meet the short- and long-term needs of both industrialized nations and LLDCs and NICs. Such a ban would provide short- and long-term protection for developing countries. At the same time, it would allow these countries to develop their own waste management systems to cope with long-term development.

As many developing and newly industrialized countries develop, they are strongly becoming waste-generating countries, if they are not already producing hazardous wastes. Thus, these countries are in the beginning of a process in which source reduction must be implemented as early as possible in order to avoid the environmentally fateful path that industrialized countries have taken. In order to design and implement industrial policies that prevent waste generation and soundly manage the resulting wastes, LLDCs and NICs need newest technologies, technical information, financial assistance, and so forth, which they can only obtain through international cooperation. Thus, in designing a global waste policy, crucial international cooperation has to be encouraged.

For industrialized countries, a total ban would in the short term cause a major problem of how to dispose of the wastes generated. This short-term pressure would, however, represent a major incentive for action towards long-term self-sufficiency. The objective to ban hazardous wastes exports to *developing countries* in order to protect the global environment can be achieved when such a ban is integrated into a larger agreement between industrialized and developing countries.[8]

In the past, some industrialized countries have adopted some guidelines for waste minimization relating to transboundary waste movements. For instance, West Germany has an agreement with France, Belgium, and the Netherlands that the authorities in the latter countries would certify that their countries did not have sufficient capacity to dispose of the waste before West Germany would consent to accept the import of waste for its disposal facility in Herfa-Neurode, one of the safest facilities available in West Germany (Kelly 1985). Nevertheless, a global export ban would certainly exercise strong pressure on industrialized countries to speed up their efforts to minimize hazardous waste generation.

---

[8]See "Evaluation of Regional Agreements" in Chapter 7.

Finally, it should be considered whether a global and complete ban would have a chance of being accepted and adopted at an international level at this time. As international negotiations have shown, many industrialized countries object to such a ban and an international waste agreement calling for such a ban still faces strong resistance.[9] Although incentives and further pressure are needed to minimize waste to the highest extent possible, realistically, a global ban on waste exports must remain a longer term objective. Policies allowing transboundary movements of hazardous wastes, with certain restrictions in order to improve the system and effectively protect the environment, only represent a short-term solution. They are examined in the next section.

RESTRICTED EXPORT POLICIES

Policies that promote free trade in transboundary movements of hazardous wastes, or that completely ban waste exports, represent two extreme policy options. A number of other options are available, some of which may be better able to take into account the special needs of the different countries. Such policies would allow hazardous waste exports only under certain circumstances that promote safer hazardous waste management. Theoretically, many scenarios for such an agreement could be drawn by establishing the criteria for restrictions and/or permissions from the combined interests and objectives of the main group of countries. In order to formulate such a policy, international negotiations were initiated by the EC, the OECD, and the UNEP, over the course of the past decade. Their approaches for a hazardous waste policy will be examined next.

The European Community

The European Community[10] became aware that a policy to control hazardous waste movements across borders was needed after 1976, when 41 drums of highly toxic waste from the Seveso clean up in Italy disappeared and could not be located by the concerned governments, until they were found in France in

---

[9]See discussion on the Basel Convention, following.

[10]The EC was founded at the Treaty of Rome in 1957. Currently, the Community consists of 12 member states: Belgium, Denmark, Federal Republic of Germany, France, Greece, Ireland, Italy, Luxembourg, the Netherlands, the United Kingdom, Spain, and Portugal. The EC has four institutions: the *EC Commission, responsible for proposing policies and administrating the Community; the Council of Ministers,* or *EC Council,* is the decision-making body and enacts legislation acting on Commission proposals; the *EC Parliament* has an advisory role and its powers lie within the control of the EC budget; and the *European Court of Justice, which interprets EC law (its rulings are binding). Environmental policies in the EC have been formulated in the four Environmental Action Programmes.*

1983. The EC concentrated its early efforts on designing a policy within the EC member states. However, the need to develop an EC policy towards non-EC member states was soon realized because large volumes of wastes were exported to non-EC member states.

Waste exports outside the EC, whether North-South, or North-East, differed essentially from the waste trade within the EC.[11] Even within the EC, the systems of hazardous waste management vary considerably, and both waste minimization and recycling of hazardous wastes is closely linked to the state of technology and the economic situation in each country. In each country, "existing environmental problems, legally binding norms and standards, problem-awareness in public authorities and the pressure of public opinion from one country to another" (STOA 1989, 1) shape and contribute to waste policies. The main consequence is a different level of technology in EC countries, resulting in cost differences for hazardous waste management—the primary reason for their export.

In the context of these widely differing national waste policies and management systems, the EC has adopted several Council Directives12 on hazardous wastes. One of the first was the Council Directive on Toxic and Dangerous Waste (78/319/EEC) (EC 1978), which provides the basis for further legislation controlling transboundary movements of hazardous wastes. Its main provisions are:

1. Definitions of "waste," "toxic and dangerous waste," as well as "disposal";[13]
2. Requirements for member states to take appropriate steps to prevent hazardous

---

[11]The EC began negotiations with the ACP countries. These are discussed.

[12]*Directives* are binding to member states with regard to the result intended, but allow flexibility in the means of achieving the result. However, member states have to implement the Directive, usually through the promulgation of national legislation or the amendment of existing laws. The Directives generally specify a deadline by which governments must implement the rules. However, these deadlines, as the case in hazardous waste export Directives, are often missed. Besides Directives, legal acts agreed on by the Council may take the form of "*regulations*," which are binding in their entirety, applying directly to all citizens in every member state. For further information, see European Parliament 1984.

[13]"Waste" means any substance or object that the holder disposes of or is required to dispose of pursuant to the provisions of national law in force (Article 1 (a)); "toxic and dangerous wastes" means any waste containing or contaminated by the substances or materials listed in the Annex to this Directive of such a nature, in such quantities or in such concentrations as to *constitute a risk to health or the environment*(Article 1 (b)); "disposal" means 1) the collection, sorting, carriage, and treatment of toxic and dangerous waste, as well as its storage and tipping above or under ground, 2) the transformation operations necessary for its recovery, reuse, or recycling (Article 1 (c)). The Annex provides a list of 27 "toxic or dangerous substances and materials selected as requiring priority consideration." See Article 1 and Annex (European Communities 1978).

waste generation[14] and take measures to dispose hazardous wastes in an environmentally safe manner.[15]

3. Establishment of an authority for planning, organization, authorization, and supervision of hazardous waste operation,[16] ensuring the labelling and data collection of hazardous wastes (Art. 7), issuance of permits for storage, treatment and disposal facilities,[17] and inspection of the HWMFs;[18]

4. Application of the Polluter Pays Principle (PPP);[19] and

5. The introduction of national plans for hazardous waste management by each member state (Art. 12), as well as a manifest system for hazardous waste transports (Art. 14), and submission of a country report on hazardous waste disposal, every three years (Art. 16).[20]

One of the main objectives of Single European Act (SEA) is to harmonize member states' laws. But in order to protect higher waste management standards of some member states, the Directive includes in Article 8, a provision giving each member state the right to take more stringent measures than required by the 1978 Directive.[21] While the need for more control and monitoring of hazardous waste exports have become more urgent, the legislative initiatives of the EC to control transboundary waste movements continue to be built on existing Directives. The important provisions of the latter directive that supersedes the former are as follows.

---

[14]"Member States shall take the necessary steps to encourage, as a matter of priority, the prevention of toxic and dangerous waste, its processing and recycling, the extraction of raw materials and possibly of energy therefrom and any other process for the re-use of such waste," Article 4 (European Communities 1978).

[15]In particular, without risk to water, air, soil, plants, or animals, without causing a nuisance thorough noise or odors, and without adversely affecting the countryside or places of special interest (Article 5) (European Communities 1978).

[16]"Member States shall designate or establish the competent authority or authorities to be responsible, in a given area, for the planning, organization, authorization and supervision of operations for the disposal of toxic and dangerous waste," Article 6 (European Communities 1978).

[17]"Installations, establishments or undertakings which carry out the storage, treatment and/or deposit of toxic and dangerous waste must obtain a permit from the competent authorities," Article 9 (European Communities 1978).

[18]Any of the installations of Article 9 shall be subject to inspection and supervisions by the competent authorities (Article 15) (European Communities 1978).

[19]"In accordance with the 'polluter pays' principle, the costs of disposing...less any proceedings from treating the waste, shall be borne by:-the holder who has waste handled by a waste collector or by an installation, establishment or undertaking refered to in Article 9(1); and/or-the previous holders or the producer of the product from which the waste came" (Article 11 (1)) (European Communities 1978).

[20]Among others, the Commission of the European Communities is required to report to the Council and to the European Parliament on the application of this Directive (see Article 12=16, in particular Article 16), (European Communities 1978).

[21]"Member States may at any time take more stringent measures with regard to toxic and dangerous waste than those provided for in this Directive" (Article 8) (European Communities 1978).

Council Directives on Transfrontier Shipments
of Hazardous Waste
Alarmed over the growing volume of hazardous waste exports and the differences among the procedures applying to the supervision and control of transfrontier shipments of hazardous wastes, the Council adopted in 1984 the first "Directive on the Supervision and Control within the European Community of the Transfrontier Shipment of Hazardous Waste" (European Communities 1984a, 31–41). This Directive was amended in 1986 with Directive 86/279/EEC,[22] which represents a strengthening of the previous Directive.

Definitions
In order to provide a common interpretation of the activities outlined in the Directive, Article 2 provides an extended list of definitions with regard to the legal and technical scope of "disposal," "competent authority," and "hazardous waste." According to the 1978 Directive, each member state defined the specific quantities or concentrations of hazardous wastes that constitute a risk to the environment (European Communities 1978, 44). As a result, the list of hazardous substances in the member states differed. Instead of the definition "toxic and dangerous waste," established in the previous Directive 78/319/EEC, the new Directive used the term "hazardous waste" and brought some improvement, in that it established a core list of 27 substances considered *hazardous* in the EC. Besides this core list, there is, unfortunately, no uniformity among member states because each member state still can add substances it considers *hazardous* (Handley 1989). Whereas PCBs were added to the list, chlorinated and organic solvents, as well as radioactive wastes, were not included. In addition, in order to come to a uniform EC-wide interpretation of the Directives, the meanings of "disposal...to protect the quality of the environment" (Art. 11) and "adequate technical capacity for the disposal" (Art. 3) need to be clearly defined. The 1984 Directive did not provide such a clear definition.

Notification and Consent
Anyone who intends to export hazardous wastes "shall notify the competent authority of the Member State" (European Communities 1986, 13) of the destination or transit country, with a unified "consignment note." The notification must contain information about the waste itself, the provisions for routes and insurance against damage to third parties, the measures to ensure safe transport, and the existence of a contractual agreement with the importer, who "should possess the adequate technical capacity for the disposal of the waste" (European Communities

---

[22]The amendments adopted replace Articles 3, 4, 5, 7, and 17 of Directive 84/631/EEC (European Communities 1986, 13=15).

1986, 13-14). Furthermore, transfrontier movements are not allowed to depart before all states involved have acknowledged receipt of the notification. This acknowledgment implies a consent if no objections are raised. The 1986 amendment expands this prior informed consent procedure to non-EC countries (Art. 4). The countries of import can object to waste shipments on the basis of "laws and regulations relating to environmental protection, public...security or health protection" and can "lay down...conditions in respect of the shipment of waste in their national territory" (European Communities 1986, 14). These conditions "may not be more stringent than those laid down in respect to similar shipments effected wholly within the Member State" (European Communities 1986, 14-15).

Information and Manifest
In order to make a sound and informed decision, the importing country is allowed to require additional information before it agrees to an export scheme (Art. 5).[23] This is important in light of the fact that a single acknowledgment may cover several shipments over a period of up to one year. The manifest, or consignment note, which is required to accompany each shipment, covers 23 pieces of information, and a copy of the manifest is retained by all undertakings involved in the shipment (Art. 6) (European Communities 1984a). Transfrontier shipments must further comply with a number of conditions, such as proper packaging, labelling, emergency instructions, and so forth. (Art. 8) (European Communities 1984a). In order to give the Commission of the EC the possibility to control the activities of its member states, Article 13 requires a biannual report on the implementation of the Directive, including information about major accidents or accident hazards involving waste shipments, any significant irregularities in waste exports, and data on the type and quantities of hazardous waste generated and traded.[24]

Liability and Insurance
The Directive incorporates the PPP, which only applies, however, to the "cost of implementing the notification and supervision procedure, including the necessary analyses and controls" (European Communities 1984a, 35), and thus does not cover damage that might result from transport and disposal of the waste. Environmental protection is covered in Article 11. According to this article and in accordance with other EC Directives, the producer of the hazardous waste "shall

---

[23]"The competent authorities of the Member State...may make their agreement to the use of this general notification procedure subject to the supply of certain information, such as the exact quantities or periodical lists of waste to be shipped," Article 5 (2) (European Communities 1986, 14-15).

[24]"Every two years, and for the first time on 1 October 1987, Member States shall forward to the Commission reports on the implementation of this Directive and on the situation with regard to transfrontier shipments concerning their respective territories" (European Communities 1984).

take all necessary steps to dispose of or arrange for the disposal of the waste so as to protect the quality of the environment."[25] The only reference to insurance is made in Article 3, which requires that the notification must include information about "the provisions made for routes and insurance against damage to third parties" (European Communities 1986, 14), but does not spell out necessary details for an insurance requirement. Neither the 1984 nor the 1986 Directive directly deal with the issue of liability in regards to transport or final disposal of the exported waste. However, in Article 11, the Directive suggests that a producer liability is intended, as the Council shall "determine not later than 30 September 1988 the conditions for implementing the civil liability of the producer in the case of damage or that of any other person who may be accountable for the said damage and shall also determine a system of insurance" (European Communities 1984a, 35).

Enforcement

EC law requires each member country to implement and enforce the Directives, by enacting national laws in accordance with the objectives of the Directives. The technical side of the Directive's enforcement lies with the "competent authority" and customs service in each country. The customs service is responsible for controlling and checking the consignment notes at border crossings (Art. 7), while the *competent authorities* in each country handle the notifications, issue permits, and forward agreements or objections to the country of export (European Communities 1978, 15). Neither of the two Directives refers to actions or procedures in case of noncompliance. This is especially important for hazardous waste exports to LLDCs and NICs, which have little power to enforce EC Directives. Within the EC, a member country at least has the right to appeal to the European Court of Justice, to seek compliance of member countries.

The EC Internal Market and the Role
of the European Parliament

With the adoption of the Single European Act in 1986, and the subsequent creation of the Internal Market by December 31, 1992, EC environmental policies will be affected by the implementation of the market but also by other policies.[26] Article 130 r-t of the SEA makes environmental policy an integral part of other EC policies, and its provisions can be interpreted as setting minimum requirements on

---

[25]This responsibility refers to EC Directives 75/442/EEC and 78/319/EEC (European Communities 1984, 35).

[26]Art. 100a of the SEA changes the voting rules from the previous unanimity rule, with which each member state could block decisions if the proposed decision did not have the desired level of protection for them, to the majority rule, which binds each member state to the majority decision.

environmental standards, while also giving its member states the right to adopt stricter rules. This creates the risk of differential regulation, resulting in economic distortions and competitive disadvantages (STOA 1989). The 4th EC Action Programme for the Protection of the Environment (1987–1992) lays out the details of the EC waste policy for the prevention, generation, recycling, and disposal of wastes. As indicated earlier, the differences in the member countries' hazardous waste policies are the main barriers for a common EC environment policy.

The implementation of the Internal Market anticipates the free movement of hazardous wastes, as trade barriers should be eliminated. However, given the difficulties and differences in current waste management, some countries would prefer that hazardous waste movements were left out. The recent discussions in the European Parliament, as well as in other EC institutions, indicate a twofold strategy for the EC policy on hazardous waste movements: (1) hazardous waste management should not be limited within national borders of EC countries, but should allow a regional trade within the EC; and (2) hazardous waste exports to LLDCs and NICs should be limited to certain conditions (Handley 1989). Besides, it appears that the EC takes initiatives to effectively minimize the volumes of wastes generated. For instance, a proposed amendment for the Council Directive 75/442/EEC on Waste (European Communities 1988a), submitted by the Commission in November 1988, calls on the member states to take measures to encourage "prevention, recycling and processing of waste," and requires that these measures "must give priority to recovery, re-use and recycling, taking into account of the available technology..." (European Communities 1988a).

The European Parliament has discussed the subject of transboundary movements of hazardous wastes several times. The disclosure of a hazardous waste export proposal to dump up to 3 million tons of hazardous wastes per year for a period of 5 to 10 years in Ginea Bissau further highlighted the waste export problem and caused irritations in the EC that funded a development project at a location close to the proposed dumpsite (Europe Environment Review 1988). After a debate in the European Parliament in May 1988, the Parliament unanimously passed a resolution condemning hazardous waste exports to LLDCs and NICs, demanding a ban on hazardous waste exports to the Third World and calling for immediate implementation of the 1984 and 1986 Directive by the member states (European Communities 1988b). By then, one and one-half years after the 1986 Directive was adopted, only Belgium, Denmark, the Netherlands, and Greece were abiding by the Directives and had passed national legislation (African Business 1988). Eight EC countries had not yet acted on the Directive and the former EC Commissioner for the Environment, Stanley Clinton Davis, announced during the debate that the European Commission would consider legal proceedings against member states if the situation remained unchanged (African

Business 1988). In late 1990, the Commission states that the transposition of these Directives "have been considerably delayed and has not yet been carried out in two Member States" (Commission of the EC 1990, 2).

Several written questions have been put forth by the European Parliament, dealing with different aspects of the subject. Frequently, Members of the European Parliament (MEPs) requested information of specific export schemes, which the Commission was generally not able to provide. Several questions addressed the issue of transboundary movement of radioactive waste. In February 1990, Mr. Ripa de Meana, Commissioner on the Environment, acknowledged in his answer to a written question by Mr. Francoise de Donnea, MEP, that it is necessary to amend the Community legislation. He stated that the Basel Convention, which the EC has signed, makes it necessary to amend the Community legislation in important respects and that the Commission intends to put forward proposals to this effect later in 1990 (see OJ No. C. 39, 19.2.1990, p. 20). On June 6, 1990, the Environmental Council of the the European Council made a landmark policy decision regarding waste exports. The Environoment Ministers of all 12 EC Member States agreed on language stating that the EC must become "self sufficient" in waste disposal and encouraged Member States to do the same. Besides self-sufficiency, the Council adopted the "principle of proximity." (European Communities, 1990). This principle means that "waste must be disposed of in the nearest suitable facility while making use of the most appropriate technologies to guarantee a high level of protection for the environment and public health." (European Parliament 1990, 5) As progressive as this statement is, according to recent information in the media, the current situation of waste disposal in Member countries that currently rely on waste exports are becoming even more dependent on such exports. Furthermore, the opinions of various committees of the European Parliament indicate that the European Parliament will not support this proposal. Its criticism could be summarized as follows:

1. The current draft of the proposal would still allow waste exports to non-EC countries and thus ignores the principle of self-sufficiency.
2. An new approach of dual definitions with respect to waste allows the opening of gaping recycling loopholes.
3. The principle of proximity may be an interesting concept, but there are reasonable doubts that it may at all be enforceable.

An EC-Directive on Liability

Neither the 1984 nor the 1986 directives deal directly with the issue of liability in regards to transport or final disposal of the exported waste. However, EC Directive 84/631/EWG required the Council to present a general regulation on liability, with respect to damage to the environment, by the end of September

1988. The Council was unable to meet this deadline, but in order to fulfill at least part of its task, it presented a directive on civil liability for environmental damage *caused by waste.*

The proposal for a "Council Directive on Civil Liability for Damage caused by Waste" (Commission of the EC 1989) presented by the end of 1989 includes 15 articles, with an introduction expressing a commitment to imposing civil liability for damage caused by wastes. Stressing that disparities among liability laws for waste in the Member States of the EC could lead to artificial patterns of investment, such a situation would also distort competition and result in differences in the level of protection of health, property, and the environment (Commission of the EC 1989). It further stresses that

> civil liability in this field should not be limited to damage and injury to the environment that occurs during transfrontier movements of hazardous wastes...[and underlines that] the principles established in 130r (2) of the [EC] Treaty that the polluter should pay and that preventive action should be taken cannot be effectively implemented in the waste management sector unless the cost of the damage or injury to the environment caused by the waste is reflected in the cost of the goods or services that give rise to the waste...strict liability of the producer constitutes the best solution to the problem (Commission of the EC 1989, 10).

Article 1 defines the application of this Directive, which is concerned with "civil liability for damage and injury to the environment caused by waste generated in the course of an occupational activity" (Commission of the EC 1989, 12), specifically excluding nuclear waste and waste and pollution from oil pollution damage.[27] Article 2 defines the terms "producer," "waste," "damage," and "injury to the environment." *Damage* is defined as "damage resulting from death or physical injury" and "damage to property" (Art. 2, (1) c). "Injury to the environment" means "a significant and persistent interference in the environment caused by a modification of the physical, chemical or biological conditions of water, soil and/or air insofar as these are not considered to be damaged within the meaning of subparagraph c) ii) [damage to property]" (Commission of the EC 1989, 12). The core of the Directive is in Article 3, which states that "the producer of waste shall

---

[27]Nuclear wastes were excluded from the application of the directive, if the wastes are covered by national law based on other international convention, notably the Convention on Civil Liability in the Field of Nuclear Energy (Paris, July 29, 1960) and the complementary convention to the latter (Brussels, January 3, 1963). Oil-related waste and pollution are exempted if they are covered by national law based on the International Convention on the Establishment of an International Convention on Civil Liability for Oil Pollution Damage (Brussels, November 29, 1969) and the International Convention on the Establishment of an International Fund for Compensation for Oil Damage (Brussels, December 18, 1971). For further details, see Explanatory Memorandum (Commission of the EC 1989, 2, 12).

be liable under civil law for the damage and injury to the environment caused by the waste, irrespective of fault on his part."[28]

In case of damage caused by waste, the plaintiff may take legal action to obtain:

1. The prohibition or cessation of the act causing the damage or injury to the environment;
2. The reimbursement of expenditures arising from measures to prevent the damage or injury to the environment;
3. The reimbursement of expenditures arising from measures to compensate for damage within the meaning of its definition;
4. The restoration of the environment to its state immediately prior to the occurrence of injury to the environment or the reimbursement of expenditure incurred in connection with measures taken to this end; and
5. Indemnification for the damage (Commission of the EC 1989, 13-14).

With regards to the restoration of the environment, restoration or the reimbursement of expenditure is excluded when:

1. The costs substantially exceed the benefit arising for the environment from such restoration; and
2. Other alternative measures to the restoration of the environment may be undertaken at a substantially lower cost (Commission of the EC 1989, 14).

The same article also limits the rights of public interest groups to take legal action. As such, it requires the plaintiff "to prove the damage or injury to the environment, and show the overwhelming probability of the causal relationship between the producer's waste and the damage or, as the case may be, the injury to the environment suffered" (Commission of the EC 1989, 14). In the event that several producers are implicated, Article 5 establishes *joint and several liability*, whereas the producer is not liable when the damage is due to "force majeure" (Art. 6). On the other hand, the article states that the liability of the producer shall not be reduced or disallowed if the damage is caused both by the waste and by an act or omission of a third party, except in case of contributory negligence—that is, when "the damage is caused both by the waste and the fault of the injured party or of any person for whom the injured party is responsible" (Commission of the EC 1989, 15). Finally, Article 14 calls on the Member States to bring into force the

---

[28]This article defines the principles of strict liability and its assignment to the producer (Commission of the EC 1989, 13).

laws, regulations, and administrative provisions necessary to comply with this Directive not later than January 1, 1991.

Evaluation

The two Directives regulating EC policy on transboundary movements of hazardous wastes are an important step towards control and monitoring of such trade. However, besides containing a number of specific problems, the overall Community's approach seems to lack the strength of an effective practical solution. Above all, the Commission itself points out that the "requirements of these Directives have certainly not been met" (Commission of the EC 1988). The following limitations emerge:

1. The "EC Waste Strategy" underlying the directives does not provide adequate rules for HWMF sitings. Due to public opposition towards new HWMFs, called the NIMBY or "not in my backyard" problem, each member state may be inclined to avoid the politically difficult siting process and try to shift the political cost of new disposal facilities to other member states. In the long term, this free-rider attitude towards hazardous waste management will result in a Community-wide shortage of hazardous waste disposal services and cause interstate difficulties (Wilmowsky 1989). Furthermore, the widely differing standards for hazardous waste management and the policy to trade hazardous wastes freely within the EC will eventually direct the waste stream into those member states having the lowest disposal standards and least stringent regulations. In the short term, this will result in the degradation of environmental quality within the EC. Nevertheless, even if the EC wanted to prohibit hazardous waste exports, it appears that such Community legislation wouid infringe upon the implementation of the Internal Market for all tradeable goods, including wastes that could be recycled.
2. Despite the efforts to provide clear definitions of what constitutes "hazardous waste," the Directives lack a comprehensive list of instrumental definitions, such as "toxic" or "adequate technical capacity," resulting in a condition that no two member states have the same legal definitions, (Yakowitz 1989a). The failure to issue an EC-wide uniform list of substances considered *hazardous* brings additional confusion in the procedures for notification and the execution of the Directives. Consequently, exporters are faced with an inconsistency among the laws of the member states (Handley 1989). The right of each member state to lay down conditions for hazardous waste imports appears to be a protective instrument to ensure public health and to prevent environmental damage. However, these conditions "may not be more stringent than those laid down in respect to similar shipment" (Art. 4). If present safety regulations cannot be tightened, then the right to impose additional safety conditions are

useless. This is motivated, however, by the EC policy to adopt uniform trade practices.

3. The PPP in Article 10 refers only to costs that result from implementing the Directive. This application is entirely unsatisfactory as it (falsely) implies a principle of responsibility for the consequences of generating hazardous wastes, and excludes in its application the polluter's responsibility from "cradle to grave." The issue of insurance and liability seems to be another loophole in the Directive. In addition, the Directive does not require insurance or financial responsibility for potential damages caused by the transport or final disposal. The provision in Article 3 to inform the importing country about insurance can at best be interpreted as a chance for the importing country to learn about whether insurance coverage exists and thus have the opportunity to object to the shipment. However, any policy on hazardous waste exports without liability from "cradle to grave" will not meet the requirement of an environmentally-sound system.

4. The Directive's requirement for biannual reports containing certain information on the Directives' implementation and accident-related hazards is insufficient as a control instrument, as it appears that no information is required on the final disposal of the wastes, and whether final disposal was performed in an environmentally sound manner. Thus, the Commission cannot monitor the waste stream from the point of generation to its final disposal. Furthermore, the time-scale of two years after which the report to the Commission has to be filed appears to be very long and does not allow the Commission to promptly respond to any irregularity, weakness in enforcement, or accident.

5. The enforcement of the Directive represents a serious problem, as the implementation of many hazardous waste Directives have systematically been delayed. Belgium was condemned by the European Court of Justice in 1982 for its failure to implement four Directives on hazardous waste. Similarly, the United Kingdom was condemned in 1986 for poor implementation of toxic waste provisions. Furthermore, evidence from the past shows that there was "almost total failure effectively to implement the clauses on…reporting to the Commission" on the waste Directives (Roelants du Vivier 1988, 11). As mentioned above, the implementation of the Directives on transfrontier shipments of hazardous wastes has proved to be equally delayed and inconsistently implemented. On a practical level, it remains to be examined whether the provisions for enforcement, and in particular the responsibilities for the customs service, will prove effective. A "myriad of regulations…[and]…the difficulty of verifying that the contents of a hazardous waste shipment match the notice and consent documentation" (Handley 1989, 10178) show that current procedures pose a problem, even for trained personnel with sophisticated

equipment. Given these conditions, existing instruments to control hazardous waste exports are insufficient.

6. Article 17 of the Directive states that "waste...from non-ferrous metals which is intended for re-use, regeneration, or recycling on the basis of a contractual agreement regarding such operations shall be *exempt* from the provisions of this Directive" (European Communities 1984a, 36), if certain administrative conditions are met. This provision appears to be a serious loophole, as it allows any exporter to engage in export schemes without having to comply with the regulations of this Directive. All that is required is the (illegal) "misclassification" of hazardous wastes as "recyclable material" or as "economic goods"—an unlawful, but actually existing activity, as has been documented.

7. Furthermore, the EC Directives must also be compared with more promising proposals by environmental NGOs and the "European Alliance for the Environment," an interparliamentary group of environmentally concerned EC-Deputees, who issued a 10-point program calling, among other items, for harmonization of hazardous waste management by the member states, adoption of a black list of waste that may not be dumped, a European catalogue setting out compulsory disposal procedures for certain types of wastes, and wider civil liability for the producers of hazardous wastes (European Environment Review 1988).

8. The proposed Directive on civil liability for damage caused by waste is a laudable effort and points in the right direction, but still does not provide a uniform system of liability. First, this Directive should be seen in the context of provisions on liability with respect to damage to the *entire environment*—not only of waste—as called for in the 1984 directive. In a draft opinion, the Committee on the Environment, Public Health and Consumer Protection of the European Parliament expressed doubt whether it makes sense to advance liability provisions that cover only waste, pointing out that "a single comprehensive *regulation* would be far more effective, as it could be used to lay down general principles...and achieve consistency with regard to the legal consequences" (European Parliament 1989, 3).

Second, the directive applies only to damage generated in the course of an "occupational activity" (Art. 1). If this means that accidents like the chemical spill at the Sandoz plant in Basel, Switzerland, in 1986, which contaminated the Rhine from Basel down to the North Sea, causing severe environmental damage, are not covered by the Directive, the Directive lacks comprehensiveness and is, from an environmental point of view, unacceptable in its current form.

Finally, provisions for fincanical responsibilities are inadequate. The Commission argues that for a number of reasons, legislation on insurance is not preferable at this time.[29] Insurance is only one form to meeting financial capability to compensate damage. The Commission should, in cooperation and consultation with the member states, recommend a uniform system of requiring financial responsibility for damage.[30] Member states may wish to choose different means of compensation in order to to bring this Directive into force.

In summary, the EC Directives do not presently meet the requirements of a sound system to control and monitor hazardous waste shipments effectively. The Directives need particular improvements in the afore-mentioned areas, and effective implementation needs to be ensured. It can further be concluded that differences between EC member states and LLDCs and NICs require a double strategy in the EC hazardous waste export policy. This view is indirectly supported by the Conseil Europèen des Fèderations de L'Industrie Chemique (CEFIC),[31] which strongly supports the Internal Market as a means to promote free trade in commodities (including hazardous wastes), but at the same time states that "no exports of hazardous waste should be undertaken unless the Generator has ascertained that the waste will be...disposed of in an environmentally sound manner" (CEFIC 1989, 17). Furthermore, recent discussions in the EC have shown that such a double strategy is likely to be adopted.[32]

Organization for Economic Cooperation and Development

The Organization for Economic Cooperation and Development (OECD)[33] was the first international organization to set up a working group to analyze the issues relating to transboundary movements of hazardous waste. In order to promote

---

[29]The Explanatory Memorandum preceding the Proposal for a Council Directive spells out three reasons why legislation on insurance was not pursued in the Directive: (1) according to the Commission, insurance companies are currently against "even limited mandatory insurance cover," as the market for environmental insurance is currently expanding rapidly, (2) that a minimum ceiling for environmental insurance (which could not be less than the ECU 70 million specified by Directive 85/374/EEC) "might distort the market," and (3) that the parties liable "may prefer other means of equal effectiveness, e.g. deposit, bill of exchange, etc." See Explanatory Memorandum (Commission of EC 1989, 5).

[30]For a comprehensive analysis of a variety of means to meet financial responsibilities for pollution-caused damage, see Ashford, Moran, and Stone 1989.

[31]CEFIC represents all 15 National Chemical Federations in Europe, which account for about 30 percent of world production in chemicals.

[32]See discussion on the negotiations for the LOME IV Treaty between the EC and the ACP, which indicates a special regulation of EC waste exports to ACP countries, Chapter 7 D.(1).

[33]The OECD has 24 member states, including the 12 member states of the EEC, Australia, Austria, Canada, Finland, Iceland, Japan, New Zealand, Norway, Sweden, Switzerland, Turkey, and the United States. Yugoslavia is involved in some work, including environmental issues.

environmentally sound waste management, the OECD Environment Committee created the Waste Management Policy Group in 1974. In 1981, this group began to prepare guidelines to identify and regulate transfrontier waste movements. Since some industrialized countries, also being members of the OECD, had already adopted legislation regulating transboundary waste movements, the OECD mainly functioned to contribute to a uniform system within its member states and to develop an international regulatory policy.

In 1984, the OECD Council adopted a Decision and Recommendation[34] on Transfrontier Movements of Hazardous Waste.[35] It states that member countries "shall control the transfrontier movements of hazardous wastes [and that]...adequate and timely" (OECD 1984, 2) information should be transmitted to the importing country. Consisting of three parts, the Decision includes *General Principles, International Pre-notification and Cooperation*, and *Definitions*. The difficulty with differing definitions in regulating hazardous waste movements was solved by defining *hazardous wastes* as any waste, other than radioactive waste, that are legally defined as such by any country of export, import, and transit. This broad definition will prevent misunderstandings and contribute to a better control and monitoring provided that effective communication exists among member countries. The principles are designed to facilitate the development of harmonized policies postulating effective control over generation, transport, and disposal of hazardous waste. The Decision underlines the sovereign right of each country to manage hazardous wastes within its jurisdiction according to national policies and regulations. It establishes a permit system requiring a permit for waste exporters. It is also important that exporters reassume:

> responsibility for the proper management of its waste, including of necessary the re-importation of such waste, if arrangements for safe disposal cannot be completed (OECD 1984, 3).

Furthermore, the countries are requested to apply laws and regulations on waste exports as stringently as in case of domestic disposal. However, as the Decision reads, the responsibility of exporters to reimport the waste can only be implied if the waste shipment was prohibited in the country of import, but not if the wastes are legally being dumped in the country of import.

---

[34]The Council is the governing body of the OECD. It is composed of representatives from each member state and has legislative authority to make " Recommendations," which are nonbinding, and "Decisions," which are binding on member states.

[35]The Decision was legally binding to all member countries that signed it (23 out of 24, excluding only Australia)(OECD 1984). Unlike Decisions, which are directly binding, Recommendations are not binding and only implemented by those member countries that consider it opportune to do so.

Notification and Export to LLDCs and NICs

The Decision further requires a notification procedure (PIC) similar to the ones described earlier, and calls on member countries to prohibit a waste export scheme being carried out, if the importing country opposes it in conformity with its legislation. Moreover, Section 7 spells out that the member countries:

> should adopt the measures necessary to enable their authorities to object to or, if necessary, prohibit the entrance of a consignment of hazardous waste into their territory, for either disposal or transit, if the information provided is insufficient or inaccurate or the arrangements made for transport or disposal are not in conformity with their legislation (OECD 1984, 4).

This provision appears to be the strongest control over transboundary movements of hazardous wastes so far, and explicitly authorizes the competent authority to stop certain waste *imports*. However, it fails to grant the same power to the competent authority in case hazardous wastes are *exported*, assuming that insufficient information on environmentally sound disposal or lack of proper arrangements is the case. The complicated issue here is deciding which legislation is applicable. For imports, the competent authority clearly applies national laws and regulations (of a typical OECD country). For exports, if the laws of the importing country are applicable, many of the exports may have been in conformity with laws (since many LLDCs' and NICs' regulations for hazardous wastes are either absent or insufficient), but they may certainly not have led to environmentally sound waste disposal. This weakness is consistent with Section 3 of the General Principles, which state that the generator of hazardous wastes should "take all practicable steps to ensure that the transport and disposal of its waste be undertaken in accordance with the laws and regulations applicable in the countries concerned" (OECD 1984, 3). It needs to be emphasized that these two provisions may be appropriate for many waste trades among industrialized countries that have equally stringent regulations, but not for waste exports to LLDCs and NICs.

This OECD Decision and Recommendation does not directly deal with waste exports to LLDCs and NICs, and their provisions do not ensure environmentally sound management of the wastes in the countries of import. Although the above provisions are a constructive first step, they need to be expanded to give the competent authority of the exporting country the right also to stop waste *exports* in addition to import and transit, if it has reason to believe that the wastes will not be disposed of in an environmentally sound manner. In regard to the Decisions' implementation and enforcement, no time limit is spelled out when the Decision has to be implemented, nor does it include any measures or penalties in the event of noncompliance.

Liability and Insurance

Provisions for insurance and liability for hazardous waste exports are entirely absent in the Decision. As a result, insurance and liability requirements apply only as mandated by national laws. The transport of hazardous wastes in many OECD countries requires insurance. Liability for damage caused while transporting hazardous wastes on the high seas are covered by other international agreements, leaving the final disposal of waste in the country of import as the only location that is not yet covered by preventive arrangements in case of an accident.[36] The issue of liability and insurance is made even more complicated, as it is not clear whether insurance and liability is the responsibility of the importing country, or whether the exporter (the waste generator) bears some or all responsibility up to the final disposal of the waste. The concept of the PPP and the widely recognized need to control hazardous wastes from "cradle to grave" would suggest such comprehensive interpretation of responsibility.

In summary, while the Decision, which is binding to member states, only requires that "Member countries shall ensure control of transboundary movements of hazardous wastes" and notify the importing government, the Recommendation, which is nonbinding, only advises member countries to implement the provisions set out in the principles. Nevertheless, the Decision and Recommendation was the first legally binding international agreement to improve control and monitoring of waste movements and laid the foundation for further action.

After an OECD Conference of Environment Ministers in 1985, the OECD Council adopted a resolution on international cooperation concerning hazardous waste exports (OECD 1985). The resolution called for the development of an international regulatory system on hazardous waste movements and instructed the OECD Environment Committee to implement this resolution and to draft such an international agreement before the end of 1987, addressing the problems of (1) Definition and Classification, (2) Notification, Identification, and Control, (3) Relations with non-Member Countries, (4) Legal and Regulatory Framework, and (5) the inclusion of appropriate economic instruments in international management and movements of hazardous wastes.

Based on the 1984 Decision and Recommendation, as well as on the 1985 Resolution, the OECD Council decided in 1986 on the proposal of its Environment Committee that member countries shall:

1. Monitor and control exports of all hazardous wastes to a final destination

---

[36]The so called HNS Convention, which is not yet in force. Furthermore, the "Convention on Civil Liability for Damage Caused During Carriage of Dangerous Goods by Road, Rail and Inland Navigation Vessels" (CRTD) will cover transfrontier shipments of HWs, when being in force.

outside the OECD area and empower their competent authority to prohibit such exports in appropriate instances;

2. Apply no less strict controls on transfrontier movements of hazardous wastes involving non-Member countries than they would on movements involving only Member countries;

3. Prohibit movements of hazardous wastes to a final destination in a non-Member country without the consent of that country and the prior notification to any transit countries of the proposed movements; and

4. Prohibit movements of hazardous wastes to a non-Member country, unless the wastes are directed to an adequate disposal facility in that country (OECD 1986).

The Council further recommended that, in order to implement this Decision, Member countries should "seek to conclude bilateral or multilateral agreements with non-Member countries to which frequent exports of hazardous wastes are taking place or are foreseen to take place" (OECD 1986, 2) and apply measures to facilitate the harmonization of national policies.

This Decision and Recommendation represents a major improvement over the 1984 Decision. Most important, it deals with the problems resulting from hazardous waste exports to LLDCs and NICs and assigns extended powers to the competent authority to stop a waste movement. Not only should member states prohibit a waste shipment unless it is directed to an adequate facility (whose interpretation should leave no leeway, but mean equally sophisticated facilities as used within OECD countries), the competent authority is also allowed to prohibit a shipment if it is not satisfied with information on the shipment provided by the exporter. It includes information that the "proposed disposal operation can be performed in an environmentally sound manner" and, regarding previous shipments, that the "wastes have been...disposed of as foreseen" (OECD 1986, 3). Hopefully, "disposed as foreseen" will be interpreted to mean environmentally sound disposal according to state-of-the-art technology and the best knowledge. In order to better control and monitor waste exports, the competent authority may also take over some of the above tasks from the exporter. In sum, the 1986 Decision and Recommendation covered the main issues called for in the 1985 Council Resolution C(85) 100, which was adopted after the Environment Ministers conference. Still, the complex issue of insurance and liability was not resolved.

In cooperation with many OECD and EEC member states, the Environment Committee proposed definitions for "wastes," "hazardous wastes," and "disposal," in addition to a comprehensive classification system for hazardous wastes

(Yakowitz 1988, 7–8).[37] The third Council Decision on transfrontier movements of hazardous wastes of 1988 finally adopted the proposals on definition and classification of hazardous wastes (OECD 1988).[38]

This classification system is a key advance in avoiding misunderstandings, and misinterpretations, as all member countries use the same definitions of the code, provided the tables are translated into the languages of the member states. Furthermore, the common list of hazardous wastes includes all wastes defined as *hazardous* "in the Member country from which these wastes are exported or in the Member country into which these wastes are imported" and thus can be considered complete for any waste movement among OECD member countries. Unfortunately, this language does *not* include wastes legally defined as hazardous in non-OECD member countries, and thus represents a loophole. This could simply be closed by stating that all wastes defined as hazardous "in the [Member] country into which these wastes are imported," while leaving out the word "Member." In 1989, the OECD Council recognized and welcomed the negotiations for a global convention which was convened under the UNEP, and instructed the OECD Environment Committee to monitor the implementation of the Basel Convention (OECD 1989a,b).[39]

Evaluation
The efforts of the OECD to establish and improve the control and monitoring of transboundary movements of hazardous wastes are laudable in that they present tangible steps for member countries. Some of the shortcomings of a few provisions, such as the right of the competent authority to prevent a shipment if it believes that the management of the wastes will not only be contrary to national laws but also that the wastes would not be disposed of in an environmentally sound manner, could be improved through amending existing Decisions. To what extent further strengthening of the present Decision and Recommendation can still be achieved and agreed upon is hard to predict, but the negotiations within the OECD countries have shown some limits. For example, the United States pointed out that it could not agree to a more comprehensive definition of what constitutes hazardous wastes

---

[37]The definition of hazardous wastes subject to transboundary movements generally *excludes* radioactive wastes and neither the OECD policy nor the Basel Convention on the Control of transboundary movements of hazardous wastes included radioactive materials or wastes. Radioactive materials, including wastes, are part of other international negotiations convened by the IAEA. The IAEA general conference in 1988 has established a "technical working group of experts to develop an internationally agreed upon code governing nuclear waste trades."

[38]Tables 1 to 6 of the classification system contain code numbers that, taken together, provide a complete system of characterization of wastes, the basis of the IWIC.

[39]For a discussion on the Basel Convention, see the analysis of the Basel Convention later in this chapter.

than that contained in RCRA, since a more stringent definition could complicate and counteract U.S. waste programs (Handley 1989).

The issues of insurance and liability will still have to be addressed, even if they are the most difficult problems to solve. A hazardous waste management system will not be publicly acceptable without an assurance for compensation in case of personal or environmental damage resulting from waste management practices. If a hazardous waste policy without such a guarantee is nevertheless in place, it must be assumed that the concerns of the people who will potentially be affected are not considered. In most industrialized countries, liability is an essential part of a hazardous waste management policy. An international policy will have to provide the same safety standards and provisions for liability if it is to be accepted by the majority of countries.

United Nations Environment Programme

Before the United Nations Environment Programme (UNEP) began its work to formulate an international policy on the transboundary waste problem, other international treaties on hazardous waste management made important contributions. In particular, the "London Guidelines for the Exchange of Information on Chemicals in International Trade" (UNEP 1989a, Appendix) and its 1989 amendment, the "London Dumping Convention"[40] and "The Cairo Guidelines and Principles for the Environmentally Sound Management of Hazardous Wastes" (UNEP 1985). The London Guidelines are addressed to governments of all countries to assist them in the process of increasing chemical safety through the exchange of information on chemicals in international trade, and are aimed at enhancing the sound management of chemicals. Furthermore, they should also support governments in developing policies for environmentally sound management of hazardous wastes. Their objective and strategy is to:

> prevent, reduce and control damage, and the risk thereof, from local and international transport as well as from handling and disposal of wastes that are toxic and dangerous to human health and to the environment (UNEP 1985, 1–2).

The London Dumping Convention was adopted to control the causes of ocean pollution and to take measures to control and reduce the dumping of hazardous wastes into the ocean. Its main result was the adoption of three lists: the "black list" of substances and wastes that are entirely prohibited to be dumped, a "grey list" of wastes that are allowed to be dumped only with a "special permit," and the "white

---

[40]The "London Dumping Convention" was adopted in 1972 and became international law in 1975. Presently, 63 countries are parties to the Convention.

list" of wastes that are allowed to be dumped with a normal permit (Art. IV).[41] The Convention also allows regional agreements and established a secretariat whose functions are carried out by the International Maritime Organization (IMO) in London (Institut für Ökologische Wirtschftsforschung 1989). The Convention is of direct relevance to transboundary waste movements, since it controls another form of hazardous waste management, which indirectly effects national waste policies. The London Dumping Convention also set up an annual meeting of the parties, which decided in 1988 to significantly reduce ocean incineration by 1989, and agreed to ban ocean incineration by 1994. The 65 countries also adopted a resolution that ordered an immediate halt to the export of noxious liquid wastes by member countries for burning by countries that are not parties to the Convention. As a result, setting up ocean incineration plants off the coast of LLDCs and NICs would be prevented (Brown 1988).

The Cairo Guidelines and Principles represent another efffort by UNEP to develop guidelines for the safe management of hazardous wastes from the point of generation to their final disposal. The guidelines include specific provisions for definitions, generation and management, monitoring and control, remedial action and record keeping, safety and contingency planning, transport, liability, and compensation (UNEP 1987; UNEP 1988). They further explore in-depth the issue of transboundary movements and management of hazardous wastes, and develop constructive rules as to how governments should respond to gain control of hazardous wastes from their generation to their disposal. With regard to transboundary waste movements, the most important provision is the right to refuse to accept waste shipments from abroad if it is assumed that such wastes could not be managed in an environmentally sound manner. However, the main weakness is that all provisions are nonbinding guidelines that merely function as a code of practice. Their provisions for protection and safe disposal of hazardous waste represent recommendations to which no government is legally bound. Correspondingly, there are no provisions for their implementation and enforcement. In sum, although these Guidelines raise crucial issues and set the tone in international discussions and negotiations regarding hazardous waste movements and waste management, their character as guidelines is not sufficient for effective control and protection of the environment.

The Cairo Guidelines were adopted by the UNEP Governing Council in 1987, the same year the Council decided to proceed with the development of a global convention concerning the control of transboundary movements of hazardous

---

[41]"In accordance with the provisions of this Convention Contracting Parties shall prohibit the dumping of wastes…except as otherwise specified below: (a) The dumping of wastes or other matter listed in Annex I is prohibited [black list]; (b) The dumping of wastes or other matter listed in Annex II requires a prior special permit [grey list]; (c) The dumping of all other wastes or matter requires a prior general permit [white list]." See Article IV (1a-c) (OECD 1989a,b).

wastes. The goals for that convention were to go beyond effective monitoring and control and to contribute to environmentally sound management of hazardous wastes. Key objectives were "to prevent developing countries from becoming repositories for improperly identified and improperly managed hazardous wastes; to pinpoint what constitutes illegal traffic; and to ensure mechanisms for redress in case of illegal or inappropriate exports of hazardous wastes to developing countries" (Yakowitz 1989b, 510–11).

UNEP called for the establishment of a legal and technical working group to draft a global convention on transboundary waste movements. Altogether, this working group met six times between October 1987 and March 1989, after which the Conference of Plenipotentiaries adopted the Basel Convention on the Control of Transboundary Movements of Hazardous Wastes and their Disposal (UNEP 1989b).42 While the third meeting, held in Caracas in June 1988 with delegations of 40 countries participating, revealed substantial differences between industrialized nations and LLDCs and NICs, the group supported the idea that an agreement could be achieved on several important principles. These included the need for equal safety standards, equal controls in waste management in exporting and importing countries (provided that the importing country agrees to accept the waste), and an understanding that a recipient country must refuse a shipment if it cannot manage the waste without damaging the environment (International Environment Reporter 1988).

The Basel Convention, adopted by the Council of Plenipotentiaries in March 1989, represents in several aspects a continuation of the OECD Decisions and Recommendations. It consists of the Final Act, 9 Resolutions, several Declarations, the Preamble, and 29 Articles with 6 Annexes. Of particular interest are: (1) Definitions and General Obligations, (2) International Co-operation, (3) Institutional and Financial Arrangements, (4) Responsibilities and Liability, and (5) Implementation, Monitoring and Enforcement.

Definitions and General Obligations

The scope of the Basel Convention excludes radioactive wastes and defines "hazardous wastes" as all wastes specified in the Annex of the Convention and as any other wastes that are considered *hazardous* by the domestic legislation of the party of export, import, or transit.[43] In this regard, the Convention adopted the

---

[42]From here on *Basel Convention*. The six meetings of the working group were held in Budapest (Oct. 1987), Geneva (Feb. 1988), Caracas (June 1988), Geneva (Nov. 1988), Luxembourg (Jan. 1989), and Basel (March 1989).

[43]Article 1 defines "hazardous wastes" as "(a) Wastes that belong to any category contained in Annex I [of the Convention], unless they do not possess any of the characteristics contained in Annex III"; and (b) Wastes that are not covered under paragraph (a) but are defined as, or are considered to be, hazardous wastes by domestic legislation of the Party of export, import or transit" Article 1a-b,(UNEP 1989b).

work of the OECD working group, including the International Waste Identification Code. So far, this system is the most comprehensive in that it covers all wastes deemed hazardous by any of the parties involved. In order to ensure clear interpretation of the Convention, 21 key words were specifically defined (Art. 2). The most crucial definition is the Conventions requirement of "environmentally sound management of hazardous wastes or other wastes," which means:

> taking all practicable steps to ensure that hazardous wastes or other waste are managed in a manner which will protect human health and the environment against the adverse effects which may result from such wastes (UNEP 1989b, 42).

Rights and Responsibilities of the Parties
Besides the definitions, Article 4 states a number of general obligations on the part of signatory states. The main provisions of the Convention are:

1. Every country has the sovereign right to refuse to accept a waste shipment, and can prohibit the import of hazardous wastes (Preamble).
2. A signatory state shall not permit the export of hazardous wastes to a party that has prohibited the import of wastes (Art. 4, Sec. 1(b)).
3. Each party of the convention shall take the appropriate measures to:
   a. Reduce hazardous waste generation to a minimum;
   b. Ensure the availability of adequate disposal facilities close to the wastes generation;
   c. Prevent pollution from hazardous waste management and reduce its consequences;
   d. Reduce the practice of transboundary movements of hazardous wastes, and see that it is conducted in a manner that will protect human health and the environment; and
   e. Prohibit waste exports and/or imports if it has reason to believe that the waste in question will not be managed in an environmentally sound manner (Art. 4, Sec. 2).
4. A signatory state is not allowed to send hazardous wastes to any country that has not signed the treaty, nor to the area south of 60 South latitude (Art. 4, Sec. 6).
5. Each party shall require that hazardous wastes exported are managed in an environmentally sound manner (Art. 4, Sec. 8).
6. Parties to the Convention may export hazardous wastes, if:
   a. The country of export does not have the "technical capacity and the necessary facilities, capacity or suitable disposal sites" for safe disposal;
   b. The waste in question is a raw material for recycling or recovery (Art. 4, Sec. 9).

7. Transboundary movements of hazardous wastes are allowed to be carried out only by authorized entities and must be packaged, labelled, and transported in conformity with generally accepted international rules and standards (Art. 4, Sec. 7).

These obligations combined with the definitions represent the principal targets and further define the scope of the Convention. An exception to these obligations is found in Article 11, which provides the parties to the Convention with the right:

> to enter in bilateral, multilateral and regional agreements or arrangements regarding transboundary movements of hazardous wastes…with Parties or non-Parties provided that such agreements or arrangements do not derogate from the environmentally sound management of hazardous wastes…required by this Convention (UNEP 1989b, 56).

Thus, each party of the Convention has the right to pursue the interests of their national hazardous waste management policies. In order to control better legally sanctioned transboundary movements of hazardous wastes, the Convention pays particular attention to illegal traffic that is considered "criminal" (Art. 4, Sec. 3).

A party to the Convention is responsible for taking a waste shipment back if the export cannot be completed (Art. 8), or if the shipment was illegal (Art. 9). The provisions state:

1. When a transboundary movement of hazardous waste cannot be completed in accordance with the terms of the contract and subject to the provisions of the Convention, the State of export "shall ensure that the wastes…are taken back"[44] or find another way of disposing of it in an environmentally sound manner.
2. The reimport of hazardous wastes is also required if the transboundary movements of the wastes was *illegal*, defined as a transboundary shipment:
   a. Without notification as required by the Convention;
   b. Without the consent from the country of import;
   c. With the consent of the importing country achieved through fraud, falsification, or misrepresentation;
   d. In which the materials and the documents of the shipment did were not conform; or

---

[44]"When a transboundary movement of HW or other wastes…cannot be completed in accordance with the terms of the contract, the State of export shall ensure that the wastes in question are taken back into the State of export, by the exporter, if alternative arrangements cannot be made for their disposal in an environmentally sound manner…" Article 8 (UNEP 1989b, 52).

e. In which the wastes are disposed of in contravention of the Convention or international law.

An exception for reimport is made in case the illegal traffic of wastes is a result of conduct on the part of the importer or disposer. In that case, the State of import is responsible for the environmentally sound disposal of the wastes. In case the responsibility for the illegal traffic cannot be assigned, procedures are established for dispute resolution.

Notification, Information, and Cooperation

Monitoring and enforcement was one of the more debated issues in the negotiations on the Convention of transboundary movements of hazardous wastes. In order to monitor and enforce the Convention, the following provisions were established:

1. All persons involved in the management of hazardous wastes are required to have a national permit or otherwise be authorized to carry out shipments (Art. 4, Sec. 7 (a)).
2. The manifest system covers the waste stream from the point of the waste's generation up to the point of final disposal. Hazardous waste movements are required to have a manifest or movement document (Art. 4, Sec. 7 (c)). Each entity that handles the hazardous waste must sign a manifest either upon delivery or receipt of the waste in question. Furthermore, the disposer of the waste must send a message to the exporter and the country of export after the disposal was completed as specified in the notification. If the country of export does not receive such a notice, it shall so inform the country of import.
3. The transboundary movements of hazardous wastes requires the prior notification of countries of import and transit (Art. 6, Sec. 1) and the consent of these countries (Art. 6, Sec. 2), called prior informed consent (PIC). The Convention establishes a notification procedure similar to the one of the EC and the OECD. A country of export is required to notify the competent authority of the countries of import and transit in writing any proposed transboundary shipment of hazardous waste. Importing and transit states can then either consent to the shipment with or without conditions, deny permission, or request additional information. Commencement of the shipment is not allowed until the notifier has received *written consent* from the state of import as well as confirmation of the existence of a hazardous waste management contract between the exporter and the disposer. Article 6 further requires exporters to provide information, such as the exact quantities or periodic lists of hazardous wastes shipped. A country's consent is subject to this information. The general notification and consent may cover multiple shipments over a period of up to 12 months. In sum, exports are illegal without consent from the importing country.

Exchange of Information

The Convention establishes a *Secretariat* (Art. 13, Sec. 3), and the *Conference of the Parties* (Art. 15, Sec. 5) serves to encourage the exchange and transmission of information (Art. 10). Through the Secretariat, the parties should provide information regarding their national definition of hazardous wastes and any changes thereof, if appropriate, about any decisions to limit or ban the export of hazardous wastes. Furthermore, a *yearly* report (Art. 13, Sec. 3) regarding their transboundary movements of hazardous wastes must be submitted to the Secretariat, which compiles it into reports that it circulates. The yearly reports should contain information about:

1. The amount, category, characteristics, destination, and disposal method of exported hazardous wastes, including a list of all transit countries, and the same information regarding hazardous wastes that were imported;
2. Efforts to achieve waste minimization;
3. The measures adopted to implement the Convention;
4. The effects of hazardous waste management on human health and the environment;
5. Bilateral, multilateral, and regional agreements or arrangements;
6. Accidents occurring during transport or disposal of hazardous wastes;
7. Domestic disposal options and information; and
8. The development of technologies for waste minimization and low waste technologies.[45]

The Secretariat compiles this information to prepare reports, and conveys and circulates information covering all necessary areas of hazardous waste management. Upon request, it also helps carry out the responsibilities of the Convention and, as such, may assist in the identification of illegal traffic. Its main role, however, is to supervise and facilitate the implementation of the Convention.

The Secretariat is also responsible for the preparation of the "Conference of the Parties" called for in Article 15. The Conference, which will meet not later than one year after the entry into force of the Convention, provides another opportunity to exchange information, as well as to review and evaluate the effective implementation of the Convention. In particular, the Conference can adopt, by consensus agreement, further rules and amend the Convention. Article 15 also calls for an evaluation of the Convention's effectiveness not later than three years after

---

[45]Article 13. "Transmission of Information. 1. The Parties shall, whenever it comes to their knowledge, ensure that, in the case of an accident occurring during the transboundary movement of hazardous wastes...that those States are immediately informed." See also paragraphs 2-4. (UNEP 1989b, 58-59).

its entry into force and, if necessary, the "adoption of a complete or partial ban of transboundary movements of hazardous wastes."[46]

Provisions for Cooperation
The general tone of the Convention calls for broad cooperation among the parties and between the parties and the Secretariat. An Article on *International Co-operation* calls for cooperation in activities to achieve the prevention of illegal traffic of hazardous wastes and to improve the environmentally sound management of the wastes concerned (Art. 4, Sec. 2 (h)). In practical terms, cooperation is postulated regarding:

1. The exchange of bilateral or multilateral information in order to promote sound waste management, including harmonization of technical standards and practices;
2. The monitoring of the effects of hazardous waste movements on human health and the environment;
3. The development and implementation of new low-waste technologies and the improvement of existing technologies;
4. The transfer of waste technologies and management systems; and
5. The development of a code of practice.

To summarize the provisions for monitoring and control as well as information and cooperation, the Convention underscores the need for information exchange, in order to effectively monitor and control transboundary waste movements, and recognizes the enormous demand for cooperation in hazardous waste management of many developing and newly industrializing countries.

Funding
The Convention establishes nonbinding funding mechanisms. Article 14 calls on the parties to fund the establishment of centers for training and technology transfer, regarding the waste management and minimization on a voluntary basis, and also to provide funding on an interim basis in case of a hazardous waste-related emergency, in order to minimize damage. Another funding proposal is made in Article 10, Section 3, which calls for "appropriate means"[47] to assist LLDCs and

---

[46]"The Conference of the Parties shall undertake three years after the entry into force of this Convention…an evaluation of its effectiveness and, if deemed necessary, to consider the adoption of a complete or partial ban of transboundary movements of hazardous wastes and other wastes…" (Art. 15 (7)). See also paragraphs 1–6. (UNEP 1989b, 15–16).

[47]"The Parties shall employ appropriate means to co-operate in order to assist developing countries in the implementation of subparagraphs a, b, c, and d of paragraph 2 of Article 4." See Article 10. Paragraph 3. (UNEP 1989b, 55).

NICs in the implementation of the Convention. Furthermore, the Conference adopted, in Resolution 6, a provision that invites all Signatories of the Convention to contribute funds for the operation of the interim Secretariat.

Liability and Insurance

The provision on liability is not resolved and the Convention has not yet set binding liability provisions. Mechanisms for liability and insurance of transboundary movements of hazardous wastes are contained in two articles. Article 6, Section 11 calls on the parties to cover any transboundary waste movement with "insurance, bond or other guarantee as may be required by the State of import or any State of transit which is a Party" (UNEP 1989b, 50). The parties are also requested to cooperate in adopting a protocol of rules and procedures for liability and compensation for damage resulting from waste movements and their disposal. In light of this unresolved topic, the Conference has adopted Resolution 3, which recognizes the necessity of liability provisions and requests the Executive Director of UNEP to establish a working group of legal and technical experts to develop "elements which might be included in a protocol on liability and compensation for damage resulting from transboundary movements of hazardous wastes,"[48] and to report its results to the first meeting of the parties.

Implementation and Enforcement

While there are numerous provisions for information exchange and cooperation, there are no guidelines on implementation and enforcement. One indirect reference is made in Article 9, Section 5, which states:

> Each Party shall introduce appropriate national/domestic legislation to prevent and punish illegal traffic. The Parties shall co-operate with a view to achieving the objects of this Article (UNEP 1989b, 54).

This article can be interpreted to underscore the parties' responsibility to enforce the provision of the Convention and, to Section 5, effectively to prevent illegal traffic. Furthermore, Resolution 2 invites the Executive Director of UNEP to establish a working group of legal and technical experts to develop mechanisms for the implementation of the Convention. In sum, the question of implementation and enforcement is largely unresolved. Resolution 4 of the Final Act, which is concerned with the Convention's implementation, calls upon all states to become parties to the Convention and urges them to refrain from activities that are inconsistent with the objectives of the Convention. Other sections have the capac-

---

[48]This working group was established in early 1990 and met for the first time in Geneva, Switzerland, July 2–6, 1990. See Resolution 3. (UNEP 1989b, 8).

ity to facilitate enforcement and implementation of the Convention. For example, the Secretariat could inform other parties about violations due to its access to information, such as manifests, notifications, and yearly reports. In addition, the Conference may be able to exercise some pressure to enforce the Convention, as it could serve as an arena for parties to address violations and seek enforcement and compliance through collective action.

Dispute Resolution

The convention permits resolution of disputes between parties through negotiations or any other peaceful means (Art. 20).[49] The dispute can be submitted by mutual agreement to the International Court of Justice, or to arbitration. Annex VI of the convention outlines, in ten Articles, the procedures of the arbitration. After the parties agree, an arbitral tribunal is established, consisting of three members, two of whom represent each party. The third is designated by the two and will be the chairman (Annex VI, Art. 3). The arbitral tribunal renders its decisions in accordance with international law and the provisions of the convention (Art. 5.), and by majority vote (Art. 6.1). It may further "take all appropriate measures in order to establish the facts," (Art. 6.2) and its judgement "shall be final and binding" (Art. 10.2). Disputes concerning the interpretation of execution of the judgement may be committed to the arbitral tribunal (Art. 10.3).

Evaluation

After the last negotiation meeting in Basel, the Final Act[50] was adopted by all 116 countries attending and the EC. Up to today, 54 countries have signed the Basel Convention, and 4, Jordan, Switzerland, Saudi Arabia, and Hungary, have ratified it.[51] In order for the Convention to become international law, 20 countries have to ratify it. This low number was deliberately chosen in order to have international control over transboundary waste movements soon.

The Basel Convention represents a collaborative effort of many states to design an effective policy to control and monitor hazardous wastes. It establishes effective monitoring procedures through the requirement of notification and prior informed consent, the manifest, and the annual report, which allow certain control of a country's activities. Furthermore, the adoption of a comprehensive list of definitions, including illegal traffic and the IWIC, represent essential parts of an effective

---

[49]"In case of a dispute between Parties as to the interpretation or application of, or compliance with, this Convention or any protocol thereto, they shall seek a settlement of the dispute through negotiation or any other peaceful means of their own choice." Article 20 (1). (UNEP 1989b, 68).

[50]The Final Act is comprised of the Basel Convention and the nine Resolutions that are attached to the Convention (UNEP 1989b, 6-15).

[51]The steps to become a Party to the Convention are (1) to adopt the Convention, (2) to sign it, and (3) to ratify in the domestic parliament.

policy. However, the Convention still contains weaknesses and loopholes that need to be addressed.

Inadequte Definitions

Although "environmentally sound management of hazardous wastes" has been defined in Article 2 as "taking all practicable steps" (UNEP 1989b, 42) to manage the wastes in a manner to protect human health and the environment, it does not spell out what concrete steps are meant (Chemical & Engineering News 1989). This is very important, as Article 4 states that hazardous waste exports are *not allowed* if they will not be managed in an "environmentally sound manner" (UNEP 1989b, 45). Section 10 of the same Article refers to:

> the obligation...of States in which hazardous wastes...are generated to require that those wastes are managed in an environmentally sound manner may not under any circumstances be transferred to the States of import or transit (UNEP 1989b, 47).

This provision clearly implies responsibility of the exporting country for safe disposal abroad. Does this responsibility mean that safety standards and regulations of the country of export have to be applied in the country of import? If this is the case, how could the country of export ensure their application and enforcement? Or would such an interpretation infringe on the sovereign rights of the importing country and thus be unacceptable?

According to the Convention, each party has the sovereign right to refuse a waste shipment and entirely prohibit the import of hazardous wastes into its territory (Art. 4, Sec. 1 and Preamble). Furthermore, no party is allowed to import hazardous wastes *from* a nonparty or export hazardous wastes *to* a nonparty (Art. 4, Sec. 5). This Article appears to be a contradiction to Article 11, which allows parties to enter into bilateral or multilateral agreements with other parties or nonparties. Such agreements shall be made in the spirit of the Convention, which means that waste management under such an agreement shall not be "less environmentally sound" than under the Convention. In fact, the provision allows waste contracts between any countries.

Furthermore, since there is not yet a standard for the term "environmentally sound," its interpretation is up to the parties that may engage in long-term contracts (Wynne 1989). This loophole was admitted by an UNEP official who acknowledged that "there is nothing to stop a Third World Country from making a bilateral deal with an exporter of waste...and then doing what it wants with the waste, even dumping it into the sea" (International Environment Reporter 1988, 376). Although this scenario hopefully will be unlikely, developing countries and NGOs interpret the inclusion of bilateral and multilateral agreements as a large loophole in the Convention (Greenpeace 1989).

Resolution 8 of the Final Act explicitly authorizes the Executive Director of UNEP to set up a working group of experts to elaborate guidelines for the environmentally sound management of hazardous wastes, which will be considered by the parties at their first meeting. The success of this effect may be doubtful because, without a clear definition of this term, the Convention's provisions remain weak and allow for all sorts of interpretations until the Conference adopts a clear definition, which could take several years. Furthermore, the development of technical guidelines is likely to be a difficult negotiation as there exist large differences in the scientific community on the risks of certain disposal options such as incinerators (Ministry of Environment, Nature Protection and Nuclear Safety, FRG 1990; Friedrich 1990).

The Signatories of the Final Act also committed themselves to reduce the generation of hazardous wastes and to ensure the availability of adequate disposal facilities within their territories (Art. 4). The same Article states that waste exports are allowed only if the country of export does "not have the technical capacity" and the necessary facilities or "suitable disposal sites." Given the high environmental standards in industrialized countries, many of these countries (West Germany, United States) meet the latter of these criteria, and thereby could justify the export of hazardous wastes. Thus, there may be no incentive for waste reduction. Moreover, as the technical capacities differ widely in various countries, a disposal site might not be *suitable* because public opposition prevents its siting. The Convention does not provide a rule or standard as to what "technical capacity" and "suitable" mean. As a result, because it is difficult to judge whether a country is allowed to export, it is hardly possible to rely on this.

Monitoring, Enforcement, and Implementation

The establishment of a permit/authorization, a manifest system, PIC, and a Secretariat and Conference seem to be constructive measures that allow for the monitoring of certain provisions of the Convention. The *permit* allows each party to control waste generators and transporters within its own jurisdiction; the *manifest system* allows for monitoring of the waste shipments and collecting all relevant information as required by the manifest; the practice of *Prior Informed Consent*, which is carried out between the "competent authorities"[52] of the parties concerned, allows for monitoring of whether the country of import has been informed about the waste movement and has consented to it.

Other provisions in the Convention call for cooperation, describe the functions of the Secretariat, require "environmentally sound waste management," and man-

---

[52]"Competent authority" means one governmental authority designated by a party to be responsible for receiving and responding to notifications of transboundary movements of hazardous wastes and for receiving information related to it. See Art. 2. (UNEP 1989b).

date the adoption of "appropriate measures" to determine whether waste exports are allowed. Unfortunately, there are no further guidelines as to how the cooperation to harmonize standards and waste management practices, the transfer of technology, and the development of a code of practice could be measured or monitored. The call for cooperation is not enough. What is needed are concrete technical, educational, and financial mechanisms to enable the authorities of exporting and importing countries to monitor and enforce the provisions.

The enforcement of the Convention relies on the provisions to prohibit illegal traffic, to cooperate in all aspects of the Convention, and to seek a peaceful settlement in case of disagreements. Article 9 requires each party to enact "appropriate legislation to prevent and punish illegal traffic." This legislation will, if enacted and consistently applied, allow enforcement of the convention within the jurisdiction of each party by imposing sanctions in case of illegal practice.

The Secretariat and the Conference of the parties may have the potential to enforce the Convention if it were given the necessary authority and means to exercise sanctions. The Secretariat is designated to assist the parties, and, by transmitting complaints, could play a mediating role. The UNEP Governing Council allocated up to US$ 3 million for the period of 1990 to 1991 for the interim Secretariat and the implementation of the Basel Convention. How effective the Secretariat will be is questionable, as it appears that it has no authority to exercise power. One of the functions of the Secretariat is to issue an annual report on waste shipments. It is not spelled out what the Secretariat can do if a party does not submit a report, or if it is incomplete. "Environmentally sound management of hazardous wastes" is very difficult to carry out and even more difficult to monitor, especially in LLDCs and NICs, possibly thousands of miles away from the exporting country. Who will determine whether it is monitored, and by what standards?[53]

LLDCs and NICs may not be able to implement the Convention effectively, given their limited resources and lack of guaranteed financial assistance through the Convention. This applies also for the continued monitoring and enforcement of the Convention. To what extent the working group (to be set up according to Resolution 1) can establish a mechanism to implement the Convention remains to be seen.

Liability and Insurance
The Convention's provisions regarding liability and insurance are another major weakness. The Convention states that:

Any transboundary movement of hazardous wastes...shall be covered by insurance,

---

[53]Article 4.8 calls for the development of "technical guidelines for the environmentally sound management of waste," which shall be decided by the Parties at their first meeting (UNEP 1989b).

bond or other guarantee as may be required by the State of import or any State of transit which is a Party (UNEP 1989b, 50).

Whether this insurance covers only the transport of the wastes or the entire waste stream from "cradle to grave" is not clearly defined. Furthermore, granted that a comprehensive insurance is implied, the wording of the provision suggests that only states of import or transit that are parties to the Convention may require insurance. Thus, countries that are not parties to the Convention may not require insurance. If the Convention does not want to risk environmental harm, or to apply a double standard, one for parties and one for nonparties, this wording is unacceptable.

Up to now, nothing has been agreed to on liability, and no obligations derive from the Convention if an accident might occur in the process of disposal (*Chemical & Engineering News* 1989). Article 12, Consultation on Liability, calls on the parties to adopt, as soon as practicable, a protocol on rules and procedures for liability and compensation for damage resulting from transboundary movements of hazardous wastes. This call was further underlined by Resolution 3, *Liability*, which requested the Executive Director of UNEP to establish an Ad Hoc Working Group of Legal and Technical Experts, to develop elements for a Protocol on Liability and Compensation for Damage resulting from the Transboundary Movement and Disposal of Hazardous Wastes and other Wastes, which met for the first time in Geneva, July 2-6 1990. More than 40 countries and several United Nations bodies and specialized agencies, as well as intergovernmental and nongovernmental organizations participated.

In his introductory statement, Dr. Tolba pointed out that the four ratifications since the adoption of the Convention 16 months ago is "very disappointing," as only 20 ratifications are needed for the Convention to enter into force. Mr. Tolba said that "the Convention may not be a perfect agreement. Yet it can ensure the poor nations of the South will never again be a rubbish dump of the North....It states in unequivocal terms that illegal traffic is criminal [and] enshrines the principles of notification and prior informed consent."

After holding informal preparatory consultations with a limited number of experts, UNEP prepared the meeting of the working group by sending out questionnaires to governments on the existence, scope, and coverage of provisions for liability and compensation in bilateral treaties to which they are party, as well as the existence and nature of national legislation related to the transboundary movement and disposal of hazardous wastes. The Interim Secretariat received 32 responses from governments. Following is a short summary of the evaluation of the responses:

1. Of the 32 responses, 3 indicated the existence of bilateral agreements, of which 2 have liability clauses.

2. Nearly half of the replies affirmed the existence of national provisions for liability for interterritory disposal regulations.
3. The base for exoneration from liability is very diverse.
4. The concept of damage is extended to the environment, or is unlimited, in 4 replies.
5. In all replies, except 3, liability amounts are not capped.
6. The limits for claim submissions vary, from one year to no time limits.
7. In 2 replies, compensation mechanisms included a compensation fund.
8. The majority of replies showed recovery procedures based on definitions of "pure economic loss" for damage.
9. Liability extensions beyond national jurisdictions are present with some limitations in 5 of the 32 replies.

The provisional agenda for the meeting included the following topics: Scope of Application, Definitions, Channelling of Liability, Exoneration from Liability, Financial Limit of Liability, Time Limit of Liability, Insurance and other Financial Guarantees, The Need for a Comprehensive International Liability Regime with or without Elements of State Liability, Fund, Claims Procedure, Jurisdiction of Domestic Court, Applicable Law, and Date of Application.

The first working group session, which covered topics 1 through 6, adopted the report on the session at its last meeting.[54] The discussion revealed numerous differences in views relating to the various topics, and a number of differences still exist and will have to be reconciled.[55] The working group decided to hold a second session for deliberations on Parts II and III, dealing with international liability and procedures.

Agreement on this issue will be very difficult since unmistakable provision for liability and compensation may result in severe financial consequences for exporter and generators of waste (Ashford, Moran, and Stone 1989). For example, although the Organization of African Unity proposed a series of amendments that, among others, would have "made the countries which produced waste liable for its ultimate disposal," they "were rejected after pressure from some Western nations" (African Research Bulletin 1989, 9493).

Other Issues and Concerns
At the Final Act, 35 countries and the EC, which stated that they will sign the convention, signed a declaration that included the following:

---

[54]This report consists of the following parts: UNEP/CHW/WG.1/L.1/Rev. 1, July 3, 1990,
    UNEP/CHW/WG.1/L.1/Add.1/Rev.1, July 6, 1990.
    UNEP/CHW/WG.1/L.1/Add.2/Rev.1, July 6, 1990.

[55]These differences are often expressed in the footnotes of the report.

1. Said that the Signatories of this Declaration confirm their strong intention to dispose of wastes in the country of origin;
2. Called upon the countries who will sign the Convention to make every effort to phase-out exports and imports of wastes for reasons other than for disposal in facilities established within a framework of regional cooperation;
3. Condemned any imports and exports of waste to countries lacking the legal, administrative, and technical capacity to manage and dispose of wastes in an environmentally sound manner; and
4. Stressed the necessity of effective action to achieve the reduction of the quantity of wastes by developing and encouraging waste minimization and recycling (UNEP 1989b, 24).

This declaration appears to be of stronger wording than the actual Convention, limiting and "phas[ing] out" entirely the export of wastes to LLDCs and NICs. It allows waste movements "within a framework of regional operation" which would imply waste trade within the EEC, OECD, or between Canada and the United States. In sum, it suggests another policy option, namely regional policies.

A few other comments on the Convention should be made. For LLDCs and NICs, the issue of hazardous waste exports was much more a political and moral issue than it was an economic and scientific-technical one. Their sensitivity about their national sovereignty, damaged by numerous illegal waste dumps in their countries, caused outrage and suspicion. Combined, these issues helped to build strong solidarity among LLDCs and NICs. As the negotiations proceeded and particular items, such as effective control mechanisms and how much funding would be made available from industrialized nations, revealed large differences among the countries, "the dialogue became polarized" (Piddington 1989, 36). Furthermore, over the six sessions of negotiations on the Convention, numerous and "substantial differences" (International Environment Reporter 1988, 376), and "irreconcilable disagreements" (Wynne 1989, 137) had erupted between the industrialized and developing countries. For instance, several amendments, which would have "prevented the import of wastes to countries which did not have the same level of facilities and technology as exporting nations; and insisted on sophisticated verification procedures, including inspection of disposal sites," were proposed by the OAU and a number of Asian countries, but were "rejected after pressure from some Western nations" (African Research Bulletin 1989, 9493).

In addition, the Convention faced intense time pressure during the negotiation, which did not allow a more elaborate and balanced Convention. As early as June 1988, which was after the third meeting in Caracas and nine months before the final Conference was scheduled, a UNEP official pointed out that "time is running short" (*International Environment Reporter* 1988, 376). Given the need for more time, it is difficult to understand why the final Conference was not postponed. The last two drafting meetings:

> were fragmented into sub-committees on different topics, and plenary debating sessions were suspended. The conclusions of the unrepresentative sub-committees were deemed by the Chair to be sovereign. The result of these extraordinary procedures was a significantly weaker final version than that which had been taking shape at Luxembourg (Wynne 1989, 138).

This is likely to be one of the reasons why the members of the OAU decided not to sign the Convention until they could further discuss it at their next ordinary Ministers Conference. On the other hand, the leading industrial countries, the United States, West Germany, and the United Kingdom, withheld their signature and explained that they needed more time to examine the Convention.[56] Positive response came, however, from the chemical industry, which welcomed the Convention right after it was adopted in Basel (*Das Parlament* 1989).

International NGOs and other international organizations criticized the Convention for merely being a system to globally track and monitor transboundary movements of hazardous wastes, instead of restricting them, which was the ultimate goal of the environmental NGOs and many LLDCs and NICs (Greenpeace 1989). An official internal memorandum of a UN agency summarized:

> the adoption of the Basel Convention, felicitous as it may be, has not resolved the deep divisions that surround this subject nor has it bridged the differences in approach to the movement and disposal of hazardous wastes...[and] can at best be looked upon as a small step in the right direction (UN 1989).

On the other hand, the Convention is not a static piece of legislation, but represents an essential improvement in a dynamic process of continued review, with the possibility to adjust, strengthen, or even ban the export of hazardous wastes, as was spelled out in Article 15. Despite this chance for improvements, the Convention, as it stands, does not satisfy the main concerns of many LLDCs and NICs regarding the absence of significant provisions for technical assistance and technology transfer, a clear funding mechanism, and strict liability.

Since the Basel Convention was adopted in Basel in March 1989, 53 countries and the European Community have signed the Convention. Four countries—Hungary, Jordan, Saudi Arabia, and Switzerland—have ratified the Convention.

---

[56]In the meantime, the United States and West Germany signed the Convention.

REGIONAL POLICIES

Environmental problems that are not limited to nation states, but are of a transboundary nature affecting neighboring countries sometimes can only be solved with transnational policies, such as bilateral or regional agreements. In fact, in the past 20 years many such regional conventions and protocols have been adopted. For instance, in order to tackle the broad range of coastal pollution of the Mediterranean, the Mediterranean Action Plan (MAP) was set up and since its adoption has been duplicated in many other parts of the world. Similarly, in order to control CO_2 emissions in Europe to reduce acid rain, regional approaches were chosen. The policy of the EC to control transboundary movements of hazardous wastes, which was analyzed earlier, is in fact be seen as a regional policy. But, since its main goal was a restriction of hazardous waste exports, it was analyzed in the context of the OECD and UNEP proposals. This section will examine other regional policies to control and limit transboundary movements of hazardous wastes as they have been proposed over the past several years.

EC-ACP Negotiations and the Lomé IV Agreement

Besides the work of the EC to develop internally a policy on transboundary movements of hazardous wastes, the EC also began talks with developing countries. In light of the public attention over illegal waste dumps in LLDCs and NICs, but also as a result of the European Community's involvement in funding and supporting a development project in Africa that was at a location close to a proposed waste dump, the EC was concerned about possible environmental damage from hazardous waste dumps. The EC offered technical and financial assistance to clean up dumped wastes in developing countries if requested (Rheinischer Merkur 1988). A few European countries, especially the Netherlands, had an even stronger position and favored a ban on hazardous waste exports to LLDCs and NICs. At the same time, the outrage over illegal waste movements in LLDCs and NICs brought the issue to the fore on practically every agenda of these countries. As a result, as the renegotiations for the next Convention on Lomé IV[57] progressed, it became clear that the topic of transboundary waste movements would be part of the negotiations. The parties of the Lomé Conventions come from two groups: the member countries of the EC, and countries from Africa, the Caribbean, and the Pacific, which formed a group, the ACP countries. The ACP, which strongly

---

[57]The first Lomé Convention was signed in 1975. The purpose of the Convention was to establish special trade relations between the 9 EEC countries and 46 ACP countries. Lomé I was followed by Lome II (1979) and Lomé III (1984), which was signed by 66 countries. The Lomé Conventions formulated specific objectives, principles, and guidelines for cooperation in agriculture, industry, trade, emergency aid, and investment, as well as financial and technical cooperation.

supported a decision by the OAU which had condemned hazardous waste exports, stressed during the Lomé IV negotiations that the "matter is not negotiable" (African Business 1989, 14), indicating that toxic waste exports to ACP countries could not be accepted. During these early 1989 meetings, the EC preferred rules that would allow the transfer of hazardous waste if the waste is accompanied by hazardous waste-management and -disposal technology. The EC opposed a complete ban on the ground that such a total ban of waste exports would "offer insufficient guarantees" to bring this practice to a halt (ACP 1989). Nevertheless, stressing the "exacerbation of the debt problem and further…deterioration of the environment," the joint ACP-EEC assembly adopted unanimously a resolution on the environment that recommended that:

> all movements of toxic and nuclear waste from the EEC to the ACP regions, including those via third countries, should be prohibited and that research should be undertaken with a view both to introducing effective and ecologically sound waste management and to finding substitutes for the products and processes which result in dangerous waste (ACP-EEC Joint Assembly 1989, 24).

A few weeks later, at a meeting of the ACP-Council of Ministers,[58] the Ministers adopted a resolution in which they (1) reaffirmed the ACP position with regard to a total ban of hazardous waste imports, (2) welcomed the Community's assurance to the effect that it will respect that ban, (3) insisted that the EC and its member states should adopt a total ban on hazardous waste exports to ACP countries, and (4) demanded a specific commitment from the Community and its member states to that effect and that it be inscribed in the next Lomé Convention (ACP 1989).

By mid-1989, the negotiations between the ACP and the EC had progressed to the point that the EC agreed in principle to a ban on all movements of toxic and nuclear waste from the EC to the ACP, with one notable exception. Under this exception, a hazardous waste export may be allowed on a case-by-case basis, provided that the ACP country for which a hazardous waste shipment is proposed has appropriate disposal facilities. The procedures for allowing an exception on a case-by-case basis were to be decided later. Furthermore, according to the EC proposal, ACP countries would also have to ban the import of hazardous wastes from any other country besides the EC. The EC was mainly concerned that a total ban might cause economic disadvantages and distort the competitiveness for their member countries. This would be the case if EC countries would have to manage their wastes according to more stringent standards with highly expensive waste

---

[58]The ACP Council of Ministers is the supreme organ of the ACP group; it defines the outlines of the objectives for the ACP group and also the Lomé negotiations.

management technology, compared to the other OCED countries, which could continue waste exports as a cheaper waste management option and thus have an economic advantage.

The case-by-case exemption procedures would probably be very difficult to define. It can also be anticipated that the procedures may present a difficult technical and bureaucratic problem. Besides the need for definitions on rules and standards as to when a "case" is to be exempt, a whole range of unanswered questions can be summarized in *who* decides *what* are "appropriate treatment plants" and *where*, a precondition for any exemption. The outcome of these definitions and procedures for case exemptions is not clear, and their implementation and enforcement would probably be very difficult. Finally, this case-by-case policy was given up by the EC.

The Lomé IV convention (ACP-EEC 1990) was finally signed by 68 ACP countries and the EC in Lomé on December 15, 1989. The final version recognizes the increasing environmental problems in ACP countries and responds by giving *environmental protection* its own title and listing it as the first area of ACP-EEC cooperation. The importance of environmental protection is thus manifested in ACP and EC efforts to ensure that economic and social development is based on a sustainable balance between economic objectives, management of natural resources, and enhancement of human resources. The increasing interdependence in this sphere requires mutual efforts to find solutions to such global problems as transboundary movements of hazardous wastes (Commission of the European Communities 1989). In order to solve such global problems, the convention encourages the adoption of measures that fall outside the usual scope of development aid.

In only two Articles, and at the insistence of the ACP countries, the treaty establishes two major policies, namely that "the Community [EC] shall prohibit all direct or indirect export of such waste to the ACP States" (ACP-EEC 1990, 37) and that at the same time "the ACP States shall prohibit the direct or indirect import into their territory of such waste from the Community [EC] or from any other countries" (ACP-EEC 1990, 37). These provisions represent a *factual ban* of any transboundary waste movements from the EC to the ACP States, as well as from any other country to ACP countries. However, in order to allow the processing of hazardous wastes in a country with more sophisticated processing technology, the provisions "do not prevent a Member State to which an ACP State has chosen to export waste for processing from returning the processed waste to the ACP State of origin" (ACP-EEC 1990, 37). The inclusion of radioactive wastes in the Lomé Convention is a major improvement over the Basel Convention, in which radioactive wastes were excluded. Finally, the parties emphasize the importance of efficient international cooperation in order to effectively implement their waste policy.

Implementation, Monitoring, and Enforcement

In order to implement this policy, the Contracting Parties are called to expedite the adoption of necessary internal legislation and administrative regulations. Consultations may be called if delays are encountered or upon the request of one of the parties. Strict monitoring and enforcement are called for in Paragraph 2 and are to be undertaken by the Contracting Parties (ACP-EEC 1990, 37).

Obligations for the EC

After the ratification[59] of Lomé IV according to the SEA, the Commission of the European Communities (CEC) has to take action. The currently applied directive on hazardous waste transport[60] will have to be amended to implement the new Lomé IV waste trade ban. The Commission of the European Communities is currently intending to draft a regulation[61] rather than a directive to amend the Directive 84/631/EEC.[62] Similarly, the provisions and obligations from the Basel Convention, which the EC has signed, also have to be implemented through its integration in existing or new community legislation.

Interpretation and Evaluation

The provisions on transboundary waste movements in the Lomé IV convention can be regarded as a response of the ACP States to the Basel Convention, which was considered, by many countries, as an instrument "to legalise and to facilitate the international trade in waste" (Europe 1988, 14) instead of prohibiting the waste trade between industrialized and developing countries—a demand requested by many LLDCs and NICs during the negotiations for the Basel Convention. Nevertheless, the Basel Convention solved numerous problems, such as the absence of definitions. Annex 1 and 2 of the Basel Convention were adopted as the list of wastes defined as "hazardous" under the new Lomé agreement. The Lomé convention also specifies radioactive wastes in its Annex 3 and further refers to applicable definitions and thresholds laid down in the framework of the IAEA.

---

[59]Under the terms of the SEA, the Convention first has to be approved by at least 260 of the 518 Members of the European Parliament. Thereafter, the 80 contracting parties (68 ACP States and 12 EEC States) will have a maximum period of 12 months to notify the competent EEC and ACP authorities that the Convention has been ratified according to constitutional procedures in each state. The Convention comes into force on the first day of the second month after all the EEC Member States and two-thirds or more of the ACP States have deposited their ratification documents.

[60]For a discussion of the "Directive on the Supervision and Control within the European Community of the Transfrontier Shipment of Hazardous Wastes," See Directive 84/631/EEC, (European Communities 1984, 31-41).

[61]As indicated above, EEC *regulations* are compared to directives, immediately binding on member states and do not need to be implemented through the promulgation of national legislation. See also European Communities 1984b.

[62]Directive 84/631/EEC was amended by 85/459/EEC, 86/121/EEC, 86/279/EEC, and 87/112/EEC.

To what extent will this agreement be implementable? The analysis of the Basel Convention indicated the complexity of monitoring and enforcement procedures necessary to control any waste shipments effectively. The case-by-case exceptions, originally called for by the EC, were not adopted in the final Lomé Convention. Thus, since Lomé calls for a *total ban,* complex control instruments do not need to be devised to control certain waste shipments. On the other hand, the evaluation of a *global ban* scenario argued that the successful implementation and enforcement of such a policy was considered sufficiently questionable, as doubts were raised as to whether it could be successfully negotiated. This appears to be entirely different in Lomé IV, in which the waste export ban is interwoven into a much larger agreement.

The principle agreement on a total ban between the EC and the ACP States has prospects for successful implementation and enforcement:

1. The responsibilities to implement and enforce the provisions on waste movements will be executed by the institutions that have been cooperating for more than 15 years of the three previous Lomé agreements.
2. Effective implementation and monitoring is in the interest of both Contracting Parties, as these provisions are part of a much larger agreement whose success is of mutual interest.
3. Given their resources and technical capacity, the EC countries are in a position to enforce an export ban from their countries.

Furthermore, given the interest of the EC to prevent any waste imports to ACP States, and the provisions for active cooperation as inscribed in Lomé IV, the LLDCs and NICs may be capable of preventing waste imports from other industrialized countries that may still try to export hazardous wastes to ACP States.

Global Consequences as a Result of Lomé IV

As the Lomé IV agreement represents a consistent effort to safeguard the environment of the 68 ACP countries from waste exports and thus is a desirable regional solution, the agreement does not protect nonparty LLDCs and NICs, because EC states are still allowed to export toxic wastes to other developing countries. Table 7-1 lists the non-OECD countries that lack an import ban and are likely to have difficulties managing hazardous wastes in an environmentally sound manner, but that remain countries to which hazardous wastes could be exported and dumped. In addition to possible exports from the EC, practically all OECD countries that do not belong to the EC may export hazardous wastes to non-ACP countries. The possible exports to developing countries are shown in Figure 7-1.

The agreement between the EC and ACP countries raises the following considerations. It should be in the interest of the EC to maintain a nondiscriminatory policy toward all global partners in the developing world. Thus, the EC should

TABLE 7-1.   List of non-OECD Countries that Lack Import Bans and thus Remain Subject to Legal Forms of Transboundary Waste Movements.

Afghanistan	Malta
Albania	Marshall Islands
Andorra	Mexico
Anguilla	Micronesia (Fed. States)
Argentina	Mongolia
Bahrain	Montserrat
Bangladesh	Morocco
Bermuda	Namibia
Bolivia	Naura
Brazil	Nepal
British Virgin Islands	Nicaragua
Brunei	North Korea
Bulgaria	Oman
Burma	Pakistan
China (Peoples Republic)	Palau
China (Taiwan)	Panama
Colombia	Paraguay
Costa Rica	Poland
Cuba	Puerto Rico
Cyprus	Qatar
Czechoslovakia	Romania
Ecuador	San Marino
Egypt	Saudi Arabia
El Salvador	Seyschelles
French Polynesia	Singapore
German Democratic Republic	South Africa
Gibraltar	South Korea
Honduras	Sri Lanka
Hungary	Syria
India	Thailand
Iran	Tunesia
Iraq	Turks and Caicos Islands
Israel	United Arab Emirates
Jordan	United States Virgin Islands
Kampuchea	Uruguay
Kuwait	USSR
Laos	Vietnam
Lebanon	Yemen Arab Republic
Malaysia	Yemen, P.D.R.
Maldives	

Source: Greenpeace International 1989, The International Trade in Waste: A Greenpeace Inventory.

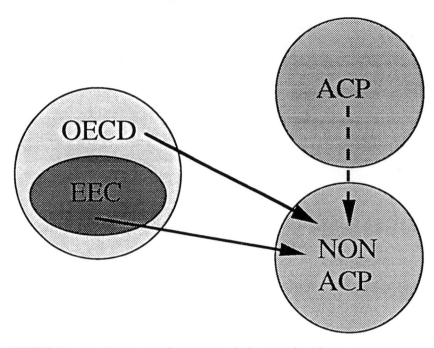

FIGURE 7-1.   Possible Exports of HWs to Developing Countries After Lomé IV.   Source: author.

adopt a uniform legislation, treating all developing and newly industrializing countries as it does the Member States of the ACP. The EC should refrain from exporting hazardous wastes to any other developing country. This would be a rational consequence and a matter of reciprocity of the ACP States agreeing not to import any hazardous wastes *from* any other countries. Legislation on waste exports outside the EC can be better and more easily administered when there is uniformity. As the development of new products and wastes progress so rapidly, law enforcement on which wastes exist and which can be exported will be much more difficult. In addition, it must be expected that national governments will outlaw waste imports, but allow certain transnational recycling schemes.[63] The issue of hazardous waste recycling becomes more pertinent as pure waste exports are banned.

A West African Policy

The African continent has encountered serious environmental degradation over the past two decades. The environmental damages range from soil erosion,

---

[63]For further discussion, see also Greenpeace 1990a.

desertification, and droughts to increasing problems stemming from fast industrialization in some countries. In addition to these problems, the dumping of toxic waste in various regions of Africa was addressed by the Heads of State and Governments of Economic Community of West African States (ECOWAS)[64] in their meeting in June 1988. The conference adopted a resolution that condemned the dumping of hazardous wastes in the territory or territorial waters of any member state. The member countries committed themselves to enact national laws to make it a criminal offence for any person or firm to engage in the dumping of hazardous wastes in ECOWAS states. Furthermore, the resolution called on the governments to take "all measures to prevent the involvement of the government or its officials, any corporate body or private citizen" (ECOWAS 1988, 1-2) in hazardous waste imports and called on the Executive Director of ECOWAS to ensure the effective establishment of a system of surveillance ("Dump-Watch") in order to ensure that the West African subregion is kept free of nuclear and industrial waste (ECOWAS 1988). Several West African countries have already enacted national legislation for waste dumping, including stiff penalties in case of such illegal activities. The Ivory Coast adopted laws that provide prison sentences of up to 20 years and fines of up to US$ 1.6 million, and Gambia has passed similar legislation setting penalties of up to US$ 830,000 and 5 to 40 years of imprisonment for hazardous waste dumping. Other countries are expected to enact equivalent laws (Vir 1989).

There are several reasons to question whether the new laws will effectively prevent the import of hazardous wastes into Africa. First, developing countries face the same substantial problems as industrialized countries do in developing a national or regional hazardous waste policy. They must solve the problems of definitions, monitoring and control, and enforcement, which, in light of their limited resources and administrative constraints, is an overwhelming task. The establishment of a "Dump-Watch" system is certainly a resolute response, but whether the member states will be in a position to implement it effectively is doubtful.

Second, enforcement of the laws may put too much of a burden on the customs service. African countries annually import millions of tons of industrial chemicals for domestic manufacturing processes or pesticides ready for use. Given these volumes of raw materials and chemicals, inspections of import cargoes, manifests, and other documents may present stern problems for the customs service. Furthermore, the already very time-consuming procedures in some countries that have forced business and trade to develop skills to circumvent these obstacles will be further extended due to technically difficult cargo

---

[64]ECOWAS was founded in 1975 by 15 West African states in order to promote cooperation and development in the member states. Today, ECOWAS has 16 members.

checks. As a result, additional procedures to control imports may deal a further blow to the bureaucracy, but may not hinder "sharp traders in the waste disposal business [who] are adept at getting around legal curbs and sloppy customs checks" (South 1988, 38).

Third, there is a notion that questions the genuine depth of the commitment of African nations to halt the waste import into their countries. As one commentator wrote, the resolutions to ban hazardous waste imports "might just turn out to be pious intentions because the potential pecuniary gains from waste dumping are a powerful temptation for many...in Africa" (Schissel 1988, 47).

To summarize the approach of West African states, the majority of these countries try to halt the waste business effectively. The approach taken appears to be a strong and consequent response. However, in order to effectively implement and enforce their policy of national legislation and "Dump Watch," West African States may have to seek international cooperation and assistance. On their own, implementation and enforcement of their strategy might not be feasible. However, besides the practical constraints, the clearly and publicly stated policy of banning imports of hazardous wastes probably has had an important political effect, both domestically and internationally.

The OAU and an African Convention

At the 25th anniversary of the Organization of African Unity (OAU),[65] the Council of Ministers of the OAU took a strong position on transboundary waste movements into Africa. The Council (1) declared that the dumping of nuclear and industrial wastes in Africa is a crime against Africa and the African people, and (2) condemned all transnational corporations and enterprises involved in hazardous waste exports to Africa, and demanded that they clean up the areas that have already been contaminated by them (OAU 1988).

This resolution also called for appropriate steps to include the item of toxic waste dumping in the 43rd Session of the UN General Assembly[66] and encouraged member states to participate in the UNEP negotiations for a global convention (Basel Convention) on this matter. At the following meeting of the OAU Council of Ministers in late February 1989, a month before the Basel Convention was to be signed, the OAU Council recalled the disagreements between African and European countries during the Euro-African Ministerial Conference on cross-border

---

[65]The OAU was founded in 1963 by 32 governments and today represents 50 countries. Its main objectives are to promote unity and solidarity of African states and promote cooperation in all fields of relevance of its member states.

[66]The UN General Assembly subsequently condemned the illegal export of hazardous wastes to developing countries.

movement of hazardous wastes[67] and expressed concern that the Basel convention was "merely aimed at the regulation or control, rather than the prohibition, of transboundary movements of hazardous wastes" and that the "narrow scope of the draft convention still permits of the illegal export of hazardous wastes...into Africa" (OAU 1989b).

At the last session before the Basel Convention was signed, the OAU was strongly represented with 41 member countries attending. But, according to their analysis of the convention, because of the loopholes in the convention, the OAU states refused to sign the Basel Convention before they could discuss further steps at their next OAU meeting. Shortly thereafter, a conference of OAU Ministers of Health proposed to establish an "African Regional Environment Surveillance Agency" (ARESA) and called upon the Secretary-General of the OAU to set up a committee of experts to draw an African Convention. The idea of a new regional policy was born (OAU 1989c).

At the 50th session of the OAU held in July 1989, the OAU Council of Ministers adopted a "Declaration on the Environment," called for an "African Year on the Environment," and set up a working group to prepare a draft African Convention and to submit its report by June 1990 (OAU 1989d). The following summit of OAU heads of state adopted a resolution in support of an African Convention, and also mandated the OAU Secretary-General to "undertake consultations for the purpose of arriving at a joint African position on the Basel Convention...to ensure the formal approval or ratification of the Basel Convention...as soon as the majority of OAU member states have signed it" (OAU 1989e).

Two years after the Basel Convention was signed, only one African country, namely Nigeria, has signed the convention. All in all, this expresses a very weak support for the Basel Convention from the African continent and is consistent with the earlier OAU positions. In the meanwhile, the first meeting of the Working Group was held in Ethiopia, December 13–16, 1989, and a draft Convention comprising of a preamble and 29 articles and annexes was adopted.

The draft African Convention, adopted at the first meeting of the working group in December 1989, aims at *banning all imports* of hazardous wastes into Africa and to that end requires that "all parties shall...prohibit the import of all hazardous wastes into Africa from non-contracting parties, whether such wastes

---

[67]The Euro-African Ministerial Conference on cross-border Movements of HWs was hosted by UNEP and the government of Senegal in Dakar (Senegal) from January 26–27 1989, with the purpose to bring together an exchange of views on the waste export matter. The European countries tried to convince African countries to participate in the on going negotiations and Convention, and to arrive at a common Euro-African position. The African countries wanted to discuss the issue on the basis of the OAU resolutions banning HW exports. The conference ended without harmonization in the issue and without a common declaration. See OAU 1989a.

are proposed for disposal, transit, or reuse/recycling" (AALCC 1990, 5)[68] and ensure that "transboundary movement of wastes *within* Africa are reduced to the minimum" (AALCC 1990, 6). Another explicit objective of the convention is to prohibit the dumping of hazardous wastes at sea, including ocean incineration (AALCC 1990). The convention also addresses the question of waste generated in the African continent. According to Article 4, Paragraph 2 (b) of the draft convention, the parties are required to commit themselves to the adoption and implementation of preventive, precautionary measures to pollution, and prevent the release into the environment of hazardous wastes by employing clean production (AALCC 1990). The same article obligates the parties to control all carriers within their "internal waters, territorial seas, exclusive economic zones and continental shelf" (AALCC 1990, 6). This definition will practically allow environmental control in export processing zones (industrial free zones), areas of concentrated environmental pollution.

General Obligations and State Responsibility

The draft African Convention stipulates five major general obligations from the contracting parties. These relate to:

1. Prohibition on import of hazardous wastes;
2. Ban on dumping of hazardous wastes at sea by all parties whether in international or territorial waters;
3. Wastes generated in Africa: commitments by each party to perform complete waste audits regularly, strict joint and several liability, generation and movements of hazardous wastes reduced to a minimum, prevention of waste generation by employing clean production, availability of adequate disposal facilities, prohibition of hazardous waste exports to countries that ban imports;
4. Transboundary movements of hazardous wastes within Africa require notification and requirements virtually identical to the Basel Convention;
5. Adoption of the Precautionary Principle (AALCC 1990; Greenpeace 1990b).

The Scope of the Draft African Convention

At least two other aspects of the African Convention are of importance. First, while the Basel Convention excluded radioactive wastes, draft Article 1, Paragraph 3 of the OAU convention includes "wastes which, as a result of being radioactive are subject to any international control system including international instruments, applying specifically to radioactive materials" (AALCC 1990, 7). Besides *radioactive wastes*, four categories of other wastes are defined in the convention,

---

[68]For an Analysis of the Draft African Convention, see Asian-African Legal Consultative Committee (AALCC) 1990.

including Annex I-III of the Basel Convention. Second, Article 5, Paragraph 3 (c) of the draft convention envisages that the parties will commit themselves to the imposition of "strict as well as joint and several liability on hazardous wastes generators in Africa" (AALCC 1990, 9). In this regard, Article 12 further stipulates the laying down of appropriate procedures and rules regarding liabilities and compensation for damages resulting from transboundary movements of hazardous wastes. According to the available information at this time, it is not clear whether these liability provisions are limited to transboundary movements of hazardous wastes only or if they also cover the other categories of waste identified in Article 1 on the scope of the Convention.

Implementation and Enforcement
In order to implement and enforce the convention's provisions, the convention provides compulsory enforcement measures with respect to the objective of prohibiting the import of all hazardous wastes into Africa. Article 4, Paragraph 2(b) states:

> Any dumping [of wastes in the parties jurisdiction] shall be deemed to be illegal. Each Party shall introduce appropriate national/domestic legislation for imposing criminal penalties on all persons subject to their jurisdiction who have planned, committed, or assisted in such illegal dumping. Such penalties shall be set sufficiently high to both punish and deter such conduct (AALCC 1990, 11). Other provisions on enforcement call on the parties to "take appropriate legal, administrative and other measures to implement and enforce the provisions of this Convention" (AALCC 1990).

Scheme of the Convention
Among other provisions, the draft convention calls for information and prior consent procedures for transboundary waste movements within Africa, similar to the Basel Convention. Other provisions call for inter-African cooperation and international cooperation and envisage in this regard further bilateral and multilateral arrangements. The draft convention also calls on the parties "to keep under continuous review and evaluation the effective implementation of the Convention" (AALCC 1990, 12).

Evaluation
The Asian-African Legal Consultative Committee (AALCC) concludes that the African Convention *does not* signify a rejection of the Basel Convention, but rather manifests an example of a multilateral, regional agreement that is provided for by Article 15 of the Basel Convention. Thus, according to the AALCC, the African Convention *complements* and *supplements* the Basel Convention, instead of replacing it (AALCC 1990). Nevertheless, the question remains as to why the member states of the OAU identify the need for their own convention. As can be

concluded from the various provisions of the draft African Convention, the answer probably lies in the shortcomings of the Basel Convention. The AALCC appraisal also concludes that the adoption of the African Convention "would not preclude" the ratification of the Basel Convention (AALCC 1990, 13-14). Given the many references to the Basel Convention, a ratification of the Basel Convention by the parties of the African Convention seems consistent.

It appears, however, that there are also negative consequences of entry into the existing agreements. For instance, the implementation and enforcement strategies of two conventions may be financially inefficient and cause bureaucratic difficulties, resulting from similar responsibilities of two separate agencies. On the other hand, the inclusion of joint and several liability will certainly have significant consequences as it incorporates a new section in the matter of international liability for injurious consequences arising out of acts not prohibited by international law (AALCC 1990). Furthermore, the provision "to keep under continuous review and evaluation the effective implementation of the Convention" appears to be much stronger than those of the Basel Convention (AALCC 1990, 12).

The debate during the Council of Ministers conference showed that there was no common position as to whether the Basel Convention should be signed. Four strategies seem to emerge from the debate: (1) the OAU signs the Basel Convention in the name of its member states, (2) the OAU drafts a convention for Africa, and then signs the Basel convention, (3) the OAU member states sign the Basel Convention as nation states, or (4) the Basel Convention will not be signed. The decisions taken by the Council and the Heads of State indicate they are inclined toward the second strategy. This strategy has several advantages. An African Convention might serve to underline the political solidarity against the "garbage imperialism" and also force single African countries that may want to consider "lucrative offers of importing waste" to comply with the African Convention. The Convention will set up mechanisms to enforce the total import ban called for in the previous OAU resolutions. Convening an African Convention and then signing the Basel Convention would also emphasize those provisions of the Basel Convention that prohibit hazardous waste exports, and at the same time allow African countries to benefit from the provisions on cooperation, technology and information transfer, and even from financial assistance, if such mechanisms can be decided upon.

To what extent the African Convention can be implemented and enforced and thus provide more protection from illegal waste imports remains to be seen after the final convention is signed and ratified. The difficulties of setting up effective hazardous waste management systems in countries with different industries, economic priorities, and many other distinctions certainly has constraints, of which some were discussed earlier. Nevertheless, an African Convention may be a means to fill the gaps and loopholes of the Basel Convention.

Evaluation of Regional Agreements

This section presented three regional policies to control, monitor, and possibly limit transboundary movements of hazardous wastes. Among these three, only the ECOWAS policy is in the phase of implementation. The opportunities and constraints that it contains have been described above, and it can be concluded that their collective regional response has had a positive impact in their overall goal to halt the import of hazardous wastes into their region. Furthermore, the ECOWAS policy must also be seen in the broader context of the OAU propositions, which are not yet complete as discussions continue. Presently, the OAU member states have scheduled a meeting on this matter at the time of this writing. As their position is clear, their final policy may include specific mechanisms, such as the African Surveillance Agency, in order to enforce the regional import ban of hazardous wastes. Similarly, like the ECOWAS approach, the regional OAU approach was successful in that it contributed to an almost unanimous and very strong position, which probably increased their political weight during the Basel Convention negotiations. The various proposals at the Basel Convention to set up working groups in order to further define a number of provisions is perhaps a direct result of the OAUs and other developing countries dissatisfaction over these provisions and their demand for modifications. The OAU might also have been inclined to delay tactically their final position in both the Basel Convention and their own African Convention in order to gain time to influence the ongoing EC-ACP negotiations on Lomé IV.

The EC-ACP negotiations can certainly be assessed as an improvement over past practices. The basic agreement to ban the export of hazardous wastes to LLDCs and NICs is laudable as it fully recognizes the fact that many of these countries are presently not in a position to manage hazard waste exports in an effective and environmentally sound manner. The EC demand to include cases for exemptions is logical from their point of view, but was reasonably rejected and could not be maintained.

OTHER DEVELOPMENTS

Besides North America, Europe, and Africa, other regions, of course, were concerned about the potential damage from wastes exports abroad and possibly managed without the necessary care. National and regional organizations of these regions responded to the increasing trade in wastes with particular policies. In addition, another option to control the transboundary waste flow was a proposal to include waste provisions into existing trade agreements. Such a response is certainly a legitimate approach, given the view of some that hazardous waste imports and exports are only a matter of trade. This section briefly presents the developments in transboundary waste policies in Latin America and Southeast Asia and finally discusses a proposal for the General Agreement on Tariffs and Trade (GATT).

The 6th Intergovernmental Regional Meeting on the Environment in Latin America and the Caribbean adopted, in March 1989, the Declaration of Brasilia (UNEP 1989c), on socioeconomic development and environmental protection. The declaration unmistakably brings the issue of uncontrolled hazardous waste exports and dumping in LLDCs and NICs, along with numerous other environmental threats, into the wider context of development. Calling for greater efforts to protect natural resources and:

> to cease practices that are highly deleterious...such as the indiscriminate and/or illegal transport and disposal of hazardous wastes and materials, and dumping in the ocean of such wastes, which places coastal areas in the whole region...in jeopardy (UNEP 1989c, Sec. 11),

the Declaration emphasizes that:

> the solution to the external debt problem and the establishment of a just and equitable New International Economic Order are imperative conditions to strengthen democracy...and to promote security and peace in the region, as well as sustainable economic and social development (UNEP 1989c, Sec. 2).

This Declaration indeed puts environmental and socioeconomic issues in their context of interdependence and urges a solution that must be within the framework of a "New International Economic Order." This proposition certainly challenges any other policy approach discussed above in that it implies that policies that do not take into account the particular socioeconomic conditions of LLDCs and NICs only try to remedy the consequences of hazardous waste exports, but do not tackle the roots of the problem. It further implies that any policy that does not consider the current economic order is likely to fail because the cause of the problem is not eliminated. However true, it is considered politically unfeasible to solve the problem resulting from uncontrolled waste exports by approaching policy options that include the reorganization of the current economic order. The principal reason for that is that it would expand the negotiations on a specific problem to a much larger and far more complex scale and scope. As a result, such an approach has a much smaller chance of being adopted.

So far, countries in the Asian and Pacific regions have not adopted a specific regional policy to control and/or limit the imports and exports of hazardous wastes. Reportedly, there were several proposals to ship hazardous wastes to Asian countries, of which many were rejected. According to the East-West Center in Honolulu, Hawaii, these countries basically apply national laws either to allow or prohibit waste imports. Countries like Malaysia, which have built modern incinerators, also depend to a certain degree on imported waste in order to operate the incinerator economically. In summary, it can be concluded that such national approaches are not sufficient, as too many loopholes exist in the current hazardous waste export policies and as many countries do not have proper national legislation.

The 1988 annual conference of the GATT[69] was also confronted with the exports of hazardous wastes. A proposal, put forth by Nigeria and supported by many other LLDCs and NICs, wanted to limit the exports of dangerous products and toxic wastes. Based on the "Tokyo Treaty," which deals with technical barriers of trade, the Nigerian proposal suggested convening an international convention to regulate the trade in hazardous waste. The proposal was rejected with the argument that the issue was not a topic of trade and should be handled within the framework of the United Nations. The controversy on this matter and the way it was dealt with caused some irritation among developing countries (Abfall 1988).

A further investigation shows, however, that Nigeria had a legitimate right to make such a proposal. The purpose of the GATT agreement is to eliminate restrictions on international trade by such measures as prohibiting import restrictions on commodities. Nevertheless, the agreement includes several exceptions under which national import restrictions can be imposed. Article XX *General Exception* states:

> Subject to the requirement that such measures are not applied in a manner which would constitute a means of arbitrary or unjustifiable discrimination between countries where the same conditions prevail, or a disguised restriction on international trade, nothing in this Agreement shall be construed to prevent the adoption or enforcement by any contracting party of measures:...(b) necessary to protect human, animal or plant life or health...(g) relating to the conservation of exhaustible natural resources of such measures are made effective on conjunction with restrictions on domestic production or consumption:....(Jackson 1969, 839–841).

Given the differing standards, regulations, and disposal practices among industrialized and developing countries, as could be observed in the waste export schemes over the past years, national restrictions on hazardous waste exports are undoubtedly justified. Trade restrictions in hazardous waste could certainly be included within the framework of the GATT agreement.

References

Abfall. 1988. Umwelt: Nord-Süd-Kontroverse um Handel mit Giftmüll. *Abfall* 47: 2.
African Business. July, 1988. Africa: The Industrial World's Dumping Ground. *African Business.*

---

[69]The GATT was set up to eliminate barriers of trade among its member states. For an explanation, see Jackson 1969.

African Business. May 1989. Lomé IV: ACP States Vow to Tackle Lomé Difficulties. *African Business.*

African, Caribbean, and Pacific Countries. February 1989. *Resolution 3./89 (XLVI) of the 46th Session of the ACP-Council of Ministers.* Brazzaville, Togo: ACP.

ACP-EEC Joint Assembly. January 27, 1989. *Resolution 4 On The Environment.* Doc. ACP-EEC 367/89/fin. 1989. Bridgetown, Barbados: ACP-EEC.

ACP-EEC. 1990. ACP-EEC *Convention of LomÉ,* ACP-CEE 2107/90, ACP/27/006/90. Brussels: EC.

African Research Bulletin. April 30, 1989. Environment. *African Research Bulletin.*

Ashford, N., S. Moran and R. Stone 1989. *The Role of Insurance and Financial Responsibility Requirements in Preventing and Compensating Damage from Environmental Risks, CTPID 89-1.* Center for Technology, Policy and Industrial Development, MIT, Boston: CTPID.

Asian-African Legal Consultative Committee. 1990. *Draft African Convention on the Control of Transboundary Movements and Disposal of Hazardous Wastes, 1989: An Appraisal.* AALCC-Doc. No. AALCC/XXIX/90/2A.

Brown, P. October 7, 1988. Global ban on toxic burning. *Guardian.*

Chemical & Engineering News. April 3, 1989. International controls on transport, disposal of wastes agreed upon. *Chemical & Engineering News.*

Commission of the European Communities. July 13, 1988. Communication from the Commission to the Council on Export of Toxic Waste. COM (88) 365 *final.* Brussels. 13. July 1988.

Commission of the European Communities. December 13, 1989. Fourth LomÉ Convention, *Information Memo.* Brussels.

Commission of the European Communities. September 17, 1989. *Com (89) 282 final—*SYN 217.

Conseil Européen Des Féderations De L'Industrie Chimique (CEFIC) 1989. Industrial Waste Management., Brussels: CEFIC.

Das Parlament. March 23, 1989. Der Giftmüll und die Dritte Welt. *Das Parlament.*

Economic Community of West African States. 1988. *Resolution A/RES.1/6/88 of the Authority of Heads of State and Government Relating to the Dumping of Nuclear and Industrial Waste.* Eleventh Session of the Authority of the Heads of State and Government. Lomé: ECOWAS.

Europe. 1988. Environment: EEE calls for provisions in the new ACP/EEC convention and a world convention on north/south transfer of dangerous waste. *Europe* 4861:14.

European Communities. 1978. *Official Journal of the European Communities.* No. L 84.

European Communities. 1984a. *Official Journal of the European Communities.* No. L 326.

European Communities. 1984b. *The European Community's Legal System.* 2nd ed. European Documentation Series. Luxembourg: Office for Official Publications of the EC.

European Communities. 1986. *Official Journal of the European Communities.* No. L 181.

European Communities 1988a. *Official Journal of the European Communities.* No. C 295.

European Communities. 1988b. Resolution on exports of toxic waste to the third world. *Official Journal of the European Communities.* No. C 167: 266–67.

European Communities. 1990. Proposal for a Council Repulation on the Supervision and

Control of Shipments of Wastes within, into, and out of the EC. COM(90) 415 final. Brussels.

European Parliament. 1984. *The European Community's Legal System*. 2nd ed. European Documentation Series. Luxembourg: Office for Official Publications of the EC.

European Parliament. 1989. *Draft Opinion*. Committee on the Environment, Public Health and Consumer Protection. European Parliament. December 14, 1989. PE 137.045.

European Parliament. 1990. Committee on Legal Affairs and citizens rights, opinions on the communication from the commission to the council and to Parliament on a community strategy for Waste Management, PE 140. 122. fin.

Europe Environment Review. 1988. Exports of dangerous wastes to Guinea Bissau and Benin. Europe Environment Review 2(2):37–38.

Frankfurter Allgemeine. January 4, 1989. Der 'Mülltourismus' bringt 20 Million Tonnen auf die Reise. *Frankfurter Allgemeine*.

Friedrich, T.A. 1990. Sondergutachten weist auf Mängel der Abfallpolitik hin; Wegwerfgesellschaft lebendiger als je zuvor *Verband der Deutschen Industrie*, Bonn.

Greenpeace. 1989. Greenpeace analysis of the Basel Convention. *Greenpeace Waste Trade Update* 2(3):3.

Greenpeace. 1990a. *The Case For Amending the EC Directive on the Tranfrontier Shipment of Hazardous Wastes to Include a Ban on Export of such Wastes to all Developing Countries*. Amsterdam: Greenpeace.

Greenpeace. 1990b. Facsimile of April 11, 1990.

Halter. 1987. Regulating Information Exchange and International Trade in Pesticides and Other Toxic Substances to Meet the Needs of Developing Countries. *Columbus Journal of Environmental Law*.

Handley F.J. 1989. Hazardous waste exports: a leak in the system of international legal controls. *Environmental Law Reporter* 89(4):10171-182.

Helfenstein, A. 1988. U.S. controls on international disposal of hazardous waste. *The International Lawyer* 22(3):775–790.

Institut für Ökologische Wirtschftsforschung (IÖW). 1989. Effizienzanalyse, insbesondere der Umweltpolitik. Berlin: IÖW.

International Environment Reporter. July 13, 1988, developed, developing countries disagree over elements of waste shipment agreement. *International Environment Reporter* (11):376.

International Environment Reporter. 1988. EC rules on waste exports often ignored; ministers disagree on tightening standards. *International Environment Reporter* (75):375–376.

Jackson, J.H. 1969. *World Trade And The Law of GATT (A Legal Analysis of the General Agreement on Tariffs and Trade)*. Indianapolis: Howard W. Sams & Co.

Kelly, M.E. 1985. International regulation of transfrontier hazardous waste shipments: a new EEC environmental directive. *Texas International Law Journal* 21(67):85-128.

Ministry of Environment, Nature Protection and Nuclear Safety. 1990. *Sondergutachten Abfallwirtschaft*, FRG.

Organization of African Unity. 1988. *Dumping Of Nuclear and Industrial Wastes In Africa*. CM/Res. 1153 (XLVII). Addis Ababa, Ethiopia: OAU.

Organization of African Unity. 1989a. *Report of the Secretary-General on the Implementation of Council of Ministers Resolutions* CM/RES.1153 (XLVIII) *Dumping of Nuclear and Industrial Wastes in Africa and* CM/RES.1199 (XLIX) Global Convention for the Control of Transboundary Movements of Hazardous Wastes. CM/1562 (L), Addis Ababa, Ethiopia: OAU.

Organization of African Unity. 1989b. *Resolution on Global Convention for the control of Transboundary Movements of Hazardous Wastes*, CM/Res. 1199 (XLIX). Addis Ababa, Ethiopia: OAU.

Organization of African Unity. 1989c. *Resolution on the Establishment of a Regional Environmental Surveillance Agency in Africa (ARESA) for Radiation and Toxic Wastes*, CM/1565 (L) Res. Kampala, Uganda: OAU.

Organization of African Unity. 1989d. *Control of Transboundary Movements of Hazardous Wastes and Their Disposal in Africa*, CM/Plen/Draft/Res. 23 (L) Rev.1. Addis Ababa, Ethiopia: OAU.

Organization of African Unity. 1989e. *Draft Resolution on Hazardous Wastes Protection of the Environment and Sustainable Development*, AHG/DRAFT. Res. 5. July 1989. Addis Ababa, Ethiopia: OAU.

Organization for Economic Cooperation and Development. February 10, 1984. C(83)180 (Final). Paris: OECD.

Organization for Economic Cooperation and Development. 1985. *Resolution of the Council on International Co-operation Concerning Transfrontier Movements of Hazardous Wastes*, C(85) 100, June 20. 1985. Paris: OECD.

Organization for Economic Cooperation and Development. 1986. *Council Decision-Recommendation on Exports of Hazardous Wastes from the OECD Area.* C(86) 64 (Final), June 11, 1986. Paris: OECD.

Organization for Economic Cooperation and Development. 1988. *Decision of the Council on Transfrontier Movements of Hazardous Wastes.* C(88) 90 (Final). *June 3,* 1988. Paris: OECD.

Organization for Economic Cooperation and Development. 1989a. *Resolution of the Council* C(89) 1 (Final) *January 30.* 1989. Paris: OECD.

Organization for Economic Cooperation and Development. 1989b. *Resolution of the Council* C(89) 112 *(Final). July 24,* 1989. Paris: OECD.

Piddington, K.W. 1989. Sovereignty and the environment. *Environment* 31(7):36.

Rheinischer Merkur. Decemder 12, 1988. Das Giftigste Metall Ist Geld: Vor Allem In Form Von Scheinen. *Rheinischer Merkur.*

Roelants du Vivier, yes F. 1988. Preventing waste and managing the burden of the past. *European Environment Review* 2(1).

Rose, E.C. 1989. Transboundary harm: hazardous waste management problems and Mexico's Maquiladoras. *The International Lawyer* 23(1):223-243.

Schissel, H. September/October 1988. The deadly trade: toxic waste dumping in africa. *Africa Report.* p. 47.

Scientific & Technological Options Assessment Programme. 1989. *Hazardous Waste Management Beyond 1992.* Technical Workshop by the Scientific and Technological Options Assessment. Luxembourg: European Parliament.

South. August 1988. The dumping grounds. *South.* p. 38.

United Nations Environment Programme. 1983. *The State of the Environment 1983*. Nairobi: UNEP.

United Nations Environment Programme. 1985. *Final Report of the Working Group of Experts on the Environmentally Sound Management of Hazardous Wastes*. UNEP/WG.122/3. December 10, 1985. Nairobi: UNEP.

United Nations Environment Programme. 1987. Cairo guidelines and principles for the environmentally sound management of hazardous wastes. *Decision 14/30 of the Governing Council of UNEP*.

United Nations Environment Programme. 1988. The United Nations Environment Programme activities in hazardous waste management. *Industry and Environment*. 11(1):32-33.

United Nations Environment Programme. 1989a. *London Guidelines for the Exchange of Information on Chemicals in International Trade*, UNEP/PIC.WG.2/4. Nairobi: UNEP.

United Nations Environment Programme. 1989b. *Basel Convention on the Control of Transboundary Movements of Hazardous Wastes and their Disposal*. UNEP/IG.80/3. March 22, 1989. Basel: UNEP.

United Nations Environment Programme. 1989c. *Declaration of Brasilia. Sixth Intergovernmental Regional Meeting on the Environment in Latin America and the Caribbean*. (UNEP/LAC-IG.VI/6. pp i-iv). Brasilia.

UN Interoffice Memorandum of March 29, 1989. Copy is in hand of author.

U.S. Congress. 1989. *International Export of U.S. Waste. Hearing before a Subcommittee of the Government Operations*, House of Representatives, 100th Congress, July 14, 1988. Washington, D.C.: Committee on Government Operations.

U.S. EPA. 1988. Agreement Between The Government of the United States of America and the Government of Canada Concerning the Transboundary Movement of Hazardous Waste. *Enforcement Strategy: Hazardous Waste Exports*. Appendix B. National Enforcement Investigation Center. Office of Enforcement and Compliance Monitoring. Denver: U.S. EPA.

Vir, A.K. 1989. Toxic trade with Africa. *Environment, Science and Technology* 23(1):25.

Wilmowsky, P. von. 1989. Transfrontier disposal of hazardous waste in the internal market: a legal perspective. Paper read at STOA-Workshop on Hazardous Waste Management Beyond 1992, 25-26 May, 1989, Scientific & Technological Options Assessment Group (STOA), European Parliament, Luxembourg.

Wynne, B. 1989. The toxic waste trade: international regulatory issues and options. *Third World Quarterly* 11(3):120-146.

Yakowitz, H. 1988. Identifying, classifying and describing hazardous wastes. *Industry and Environment* 11(1):3-9.

Yakowitz, H. 1989a. Possibilities and constraints in harmonizing national definitions of hazardous wastes. OECD *Working Paper*. 6.4.1989. W/0837M. Paris: OECD.

Yakowitz, H. 1989b. Global hazardous transfers. *Environmental Science and Technology* 23(5): 510-11.

8

Conclusions

The criteria of efficiency, equity, risk, sustainability, and feasibility in the sociopolitical context are the ultimate benchmarks of a sound policy. To what extent do the proposed policies meet these criteria?

CONTINUE CURRENT PRACTICE

To continue the practice of exporting hazardous wastes to LLDCs and NICs as has been exercised—without effective control and monitoring, and without the condition that the wastes have to be disposed of in an environmentally acceptable manner—does not represent a realistic option. The practice of transboundary movements of hazardous wastes needs to be regulated and controlled, and, in the opinion of many countries and environmental NGOs, also limited. The practice of exporting hazardous wastes without significant and effective control was the original cause of problems that now must be solved. In terms of economic benefits and the management of risks, exporting countries efficiently manage their wastes in the short run, by using the cheapest means to get rid of their wastes and risks. In the long term, however, it must be concluded that this practice will become a net economic cost as uncontrolled landfills will eventually leak and contaminants possibly enter the global food chain, posing both risks and costs, not only to societies where the wastes where dumped in the first place, but also to societies in the Northern Hemisphere.

Furthermore, even if LLDCs and NICs voluntarily engage in hazardous waste import schemes, seemingly in order to solve their debt problems, the past practice could still not be considered as ethical because many LLDCs and NICs are indirectly forced into such agreements as it appears to them that no other solutions exist to solve their problems. The fact that the export of hazardous wastes represents a transfer of potential risks abroad while all the benefits of the waste in question, either as a used product or a side product of another consumer good, remain in the country of generation further represents an unethical practice. These countries accept hazardous waste imports because of their economic situation and to encourage the continuance is to encourage exploitation of their condition.

Finally, given the above reasons and consequences, a policy of continued exports without controls does not meet the criteria of sustainability. Not only does such a practice represent a threat to natural resources and public health, one of the implications of this policy would be that the generation of hazardous wastes would practically not be limited or reduced. In contrast, such a policy represents an attractive option for waste generators, to dispose of wastes that they can hardly dispose of anymore domestically due to increasing regulatory constraints. Furthermore, the cheap option of waste exports may possibly lead to increased waste generation. This clearly does not contribute to the reduction of environmental risk and to conservation of depletable resources. Thus, it is not sustainable.

BILATERAL AGREEMENTS

The two bilateral agreements that have been examined reflect to some extent effective policies to control transboundary waste movements. Furthermore, the U.S. agreement with Mexico also limits the export of hazardous wastes to wastes that can be recycled. Given the technological capacity of Canada and its developed regulatory system, the bilateral agreement between the United States and Canada could be assessed as an economic advantage for both countries, managing the risks according to state of the art technology. Given the similar sociopolitical and economic conditions in both countries, the agreement can be interpreted as a trade agreement implying an equitable bargain with optimal economic outcome. The relatively high environmental regulations and standards in these two countries, as well as their institutional capacity, require and allow proper management of the wastes and provide that the policy can effectively be implemented and enforced.

Whether or not the U.S.-Canadian agreement results in an environmentally sustainable practice, is questionable. The import of wastes into Canada is certainly a result of economies of scale in the operation of HWMFs. The operation furthermore results in waste management, with advanced technology meeting regulatory safety and public health standards. However, the criteria of sustainability requires, among others, that waste generation be reduced to the minimum level possible, and the consumption of as few of our resources as possible in order to reduce risks to human life and to the local and global ecosystem. The notion of sustainability rejects this policy, which still results in risks that could be eliminated by waste minimization and source reduction. Since these bilateral agreements do not provide any direct or indirect incentives to reduce the use of resources and the generation of hazardous wastes, including potential risks, they are not sustainable policies.

The U.S.-Mexican agreement basically leads to the same results, provided that the agreement is effectively enforced. As shown in the analysis, the agreement could not be satisfactorily implemented because of the economic- and sociopolitical conditions in Mexico. The results of the U.S.-Mexican agreement may not be generalizable in terms of possible future agreements between industrialized countries and LLDCs and NICs. The degree to which such agreements emphasize risk reduction and equity certainly depends upon a variety of factors, such as the negotiation power of the developing country in question.

GLOBAL BAN

A complete and global ban of transboundary movements of hazardous wastes, as distinguished from a ban to export hazardous wastes to developing countries, would have a number of beneficial long-term consequences. The management of hazardous wastes would have to occur in the country where the waste was generated, usually a developed country with high technology industries. According to environmental regulations, political priorities, and public attention, the industry also would have to utilize its technological capacity to manage these wastes. Not only would such waste disposal represent a reasonable solution with regard to the costs of waste management, it would also lead to the greatest reduction of the risks involved, as industrialized countries have the most stringent environmental regulations. Furthermore, countries that do not benefit from generating wastes would not bear any financial or environmental costs. In sum, such a policy is desirable in the long term.

However, for the short term, there are a few reasons against a global ban. For instance, given the conditions of the countries, like Luxembourg, that produce only small amounts of hazardous wastes, the HWMFs they would have to build could not take advantage of the existing economies of scale. It also appears to be very difficult to implement and enforce such a complete ban if such a ban was called for by an international agreement like the Basel Convention, as many key countries objected to such a ban during the negotiations. Furthermore, a complete ban could only be implemented if it was agreed upon on a mutual basis, which would serve to encourage implementation and enforcement. Since on a global level, this is not yet the case, implementability and enforceability are questionable.

REGIONAL AGREEMENTS

Regional Agreements, as proposed by the EC, OAU, and ECOWAS, offer a variety of advantages for the member countries, but they cannot control and solve the problem of global transboundary waste movements. To briefly portray the problem, the EC has adopted a policy of waste exports and imports primarily within its member states. Although the policy was later amended to address the problems resulting from waste exports to LLDCs and NICs, it does not directly deal with countries like Switzerland, Austria, Sweden, and Norway, which are not EC member countries, but which may be both potential exporters or importers of hazardous wastes. As a result, the EC policy is certainly incomplete in that it does not provide a coherent European-wide policy.

As was shown in the section for regional agreements, the strength of a regional agreement is to address particular regional concerns, as in the OAU and ECOWAS policies, in order to provide a complementary measure of security. On a practical level, however, all regional agreements are successful only if they are integrated into the framework of a larger international agreement, such as the Basel Convention, or in a trade agreement, such as Lomé IV. If this is not the case, regional agreements are difficult to implement and to enforce. A distinction also needs to be made between the EC policy, which is implementable within the provisions for enforcement available in the EC-treaty, and the regional agreements of the OAU and ECOWAS, whose implementation and enforcement are constrained. However, it can be concluded, taking into account the differences in implementation and enforcement, that regional policies can be considered as economically efficient, regionally balanced, and adequate in managing risks within the region. Their sustainability is questionable, as the assumptions suggest that the policies support the status quo of hazardous waste generation, rather than clearly indicate the need for change in hazardous waste policies to develop industries with less or no waste.

BASEL CONVENTION

Assuming the Basel Convention will be ratified by the major exporting countries, and further assuming that the provisions calling for and ensuring environmentally sound management of hazardous wastes, including those of liability and insurance, will be clearly defined, implemented, and enforced, the Basel Convention could be considered economically beneficial. Wastes would be allowed to be exported to other countries with equally sophisticated facilities and would be very restricted to countries without appropriate disposal facilities. Under these circumstances, the risks involved in hazardous waste management will also be managed in a satisfactory manner. Furthermore, if countries with developing economies have HWMFs meeting the standards of those in industrialized countries, the Basel Convention would allow hazardous waste exports to those countries, which then would also be in a position to benefit financially from waste imports. Since most LLDCs and NICs do not now have such facilities, the Basel Convention would result in a very limited or an effective ban of exports of hazardous wastes to LLDCs and NICs, and would probably intensify trade in hazardous wastes among industrialized countries in the short term. The Convention could thus be considered equitable in the sense that it would not result in a burden for countries that could not handle hazardous waste. The lack or insufficiency of some critical provisions in the Convention, as it stands, will represent serious constraints and presently not allow a successful implementation and enforcement, as too many misinterpretations are possible. Furthermore, in the long run, what you may encourage is the large, wholesale building of HSMFs in developing countries by exporters and a lot of transported wastes with risks of spills.

Is the Basel Convention, as the most elaborate effort to control transboundary movements of hazardous wastes, an instrument supporting sustainable development? Some provisions in the Convention call for waste reduction, recycling, and reuse of hazardous wastes and thus underscore in which direction our societies must go in order to become sustainable. However, the Convention remains vague as to the degree to which waste should be reduced, as well as the methods of waste reduction. Nor does the Convention clearly state that the increasing generation of hazardous wastes over the past decades has contributed to environmental degradation worldwide and, as a result of this trend, has to be reversed. The intention of the Basel Convention was not to develop a policy to redirect the economic priorities and economies of many industrialized countries, which could be an ultimate consequence of a consistent sustainable hazardous waste policy. Nevertheless, the Basel Convention, convened by UNEP, one of the most respected environmental organizations, could have given a signal to the leading industrialized and hazardous waste-generating countries to reverse their unsustainable development by imposing pressures to adopt waste minimization policies, which can only result from a complete export ban.

Table 8-1 summarizes these results in a systematic manner in order to give an overview of proposed policies and the criteria applied. The table should be read as follows. A policy is assigned a plus (+) if it meets a certain criterion and a minus (-) if it does not meet that criterion. Admittedly, this is overly simple. It does, however, provide us the opportunity to compare the policies with each other. The results are presented in Table 8-1.

Although this table provides only a simplified assessment, it can be seen that none of the policies meets all the criteria. From this, it can be concluded that the adoption of one of these policies will ultimately result in trade-offs between one or several criteria and ultimately compromise one or several attributes. Furthermore, trade-offs between two criteria are difficult because the weight of each criterion differs in practice. Their importance depends on the many values, beliefs, and national objectives reflected in each particular culture and nation. The table identifies the issues of each policy that need improvement.

TABLE 8-1.   Policy Evaluation on Five Important Criteria.

Policy/Criteria	Efficiency	Risk	Equity	Sustainability	Implementability
Continue Current Practice	$(-)^a$	–	–	–	+
Bilateral	+	$(+)^b$	$(+)^b$	$(-)^b$	$(+)^b$
Agreement	+	$(-)^b$	$(-)^b$	$(-)^b$	$(+)^b$
Global Ban	+	+	+	+	–
Regional Agreement	+	+	+	–	–
Basel Convention	+	+	+	–	–

[a]Efficiency as a result of an overall economic benefit has to be distinguished between short-term and long-term efficiency. Furthermore, the results are different for exporting countries and importing countries. As evaluated above, for exporting countries, continued practice of hazardous waste exports results in economic benefits in the short term, and more likely in economic costs in the long term. For importing countries, (specifically developing countries), given their willingness to engage in such imports (also true for developed countries), waste imports were evaluated beneficial as well in the short term, and non-beneficial in the long term. Thus, the result is two (+) and two (-). It is valued overall as (-), because the purpose of this policy analysis is to identify a sustainable policy. Such a policy would also question the validity of currently used models of economic analysis (especially in their use of the discount rate) and suggest that the current practice is economically inefficient.

[b]Bilateral agreements among industrialized countries differ from agreements between industrialized and developing countries. The three (+) for risk, equity, and implementability are based rather on the U.S.-Canadian agreement. The agreement between the U.S. and Mexico would have a (-) for risk and implementability, as it could not meet these criteria.

Source: author.

The table highlights that the continuation of past practices is the worst option available, as it does not meet any of the criteria except that it is easy to implement by choosing to do nothing.

Bilateral agreements achieved a higher rating, but their overall success is unpredictable, as the agreement depends largely on which countries are involved (see b). Although bilateral agreements between industrialized countries represent a fairly good solution, large national differences in the implementation, enforcement, and compliance between industrialized and developing countries show that bilateral agreements are insufficient to control North-South (or East-West) movements of hazardous wastes. Therefore, bilateral agreements lack the comprehensiveness needed for a global policy.

By now, it should be proven that in the *long run*, only a global and complete ban of waste exports from industrialized to developing countries represent a sustainable policy. However, although such a global ban meets four desirable criteria, it produces economic inefficiencies in the short term and appears to be very difficult to implement if such a policy is not integrated in a larger development aid and trade agreement. As the Lomé IV convention shows, implementation and monitoring could be of mutual interest, and the agreement's provisions can thus be effectively enforced. The fact that such a policy causes short-term inefficiency must be considered the economic cost for a policy that promotes source reduction and thus moves us closer to a sustainable world.

Regional agreements represent an expanded version of bilateral agreements, but take into account the need for a larger scope. In that, regional agreements represent a possible solution within certain regions, such as the EC or the African continent. Regional agreements may, however, be very difficult to implement in LLDCs, and they may also hinder the development of hazardous waste management systems if they are not seeking international cooperation. It must, once again, be recognized that many LLDCs and NICs have many problems with domestic hazardous wastes. Finally, with regard to the problems of transboundary waste movements, regional policies do not provide a scheme to cover all transboundary waste movements and thus fail to provide comprehensive control on a global level. However, regional agreements, like Lomé IV, can result in a desirable regional solution. The Basel Convention closes the gap of limited scope by proposing a comprehensive approach of global scope. However, as it is formulated, it fails to solve crucial issues and is therefore not satisfactory.

In sum, all policy options examined are insufficient and unsatisfactory for the reasons outlined above. The analysis of the various policy options consequently leads us to make suggestions to improve the best available policy option—the Basel Convention. Subsequently, a set of recommendations specifically addressing existing weaknesses and issues of importance are proposed. These recommendations concern three levels: the international community, nation states, and the operational level of policy implementation.

9

Recommendations

1. REDEFINING NATIONAL SOVEREIGNTY

The transboundary nature of the dangers and hazards resulting from uncontrolled toxic waste exports and disposal requires not only international action, but suggests a redefinition of national sovereignty in environmental protection issues. In order to allow rapid international response in case of an emergency, national sovereign rights regarding the environment should be limited in favor of more extended rights of an international agency.[1] In emergency cases, which are to be defined by the Protocol, when large environmental damage can be anticipated, sovereign rights should be given up for the objective called for by Principle 24 of the United Nations Conference on the Human Environment:

> to effectively control, prevent, reduce and eliminate adverse environmental effects...in all spheres, in such a way that due account is taken of the sovereignty and interests of *all* States (International Protection of the Environment 1975, 120).

The result would be shared rights and obligations in order to respond promptly and better protect the environment. It could be argued that, not only emergency situations, but the ethical responsibility of all nations to protect the global environ-

---

[1]The need to redefine national sovereignty in environmental matters became an issue after the Chernobyl nuclear accident. National sovereignty basically prohibited active international action to prevent environmental damage timely and comprehensively, as the acceptance of offered international assistance was subject to Soviet Union approval only.

ment "requires an abandonment of frontier values" (Piddington 1989, 38). Appropriate enforcement of the Protocol may require measures such as a check system that allows international inspectors to enter countries to inspect HWMFs in order to determine whether or not the wastes can be disposed of in an "environmentally sound manner." Such measures, though, would certainly affect sovereign rights.[2]

2. IMPROVEMENTS OF THE BASEL CONVENTION

As concluded above, the most effective means to control transboundary waste movements requires a supranational approach. This can be accomplished through an international protocol. In order to make this feasible, a concerted international effort has to be made, most practically through renegotiation of the Basel Convention or by its further improvement at the first conference of the parties.

Industrialized countries and developing countries bear equal responsibility for the safe management of hazardous wastes. Developing and newly industrialized countries facing rapid economic growth and increased generation of hazardous wastes must direct their economic development within a sustainable policy. And industrialized countries, deeply involved in LLDCs and NICs through numerous economic activities, should make an extra effort to assist these countries in developing such policies. This requires political will and long-term commitment from both parties. The result will be mutually beneficial by protecting natural resources and limiting environmental damage. To give an example, in contrast to the Basel Convention, in which many critical provisions were weak or absent, industrialized countries would have to take responsibility for the wastes they generate by ensuring their environmentally safe disposal. This would involve financial assistance for appropriate HWMFs and emergency measures, as well as assuming full liability. LLDCs and NICs, on the other hand, must recognize the close link between development and the environment and consequently give environmental protection equal consideration along with economic development.

Since industrialized countries cannot be forced to participate in negotiations, we must appeal to their sense of moral responsibility for our common environment to participate in good faith in the negotiation for an improved international protocol. It must be clearly established that without their contribution, a solution is not in sight. Furthermore, the waste generating countries also must reevaluate their own waste policies. This ranges from a critical review of domestic environ-

---

[2]Currently, the IAEA has an inspection system that allows IAEA inspectors to check nuclear facilities in member states. Furthermore, The U.S. Food and Drug Administration (FDA) has for example, the right to inspect food processing facilities abroad (mainly in LLDCs) to assess their compliance with U.S. standards and regulations. These examples show that national sovereignty is already limited in some cases.

mental regulations for hazardous waste generation to economic incentives for waste minimization and source reduction.

The international Protocol should not allow bilateral agreements that are contrary in purpose or effect to what is intended by the Protocol. However, agreements in regional cooperation, such as within the EC, which is in the process of harmonizing hazardous waste management in its member states by 1993, should be encouraged. The result would be that control and regulation of transboundary movements of hazardous wastes would solely be controlled by the international Protocol. This will contribute to protect all countries concerned, especially LLDCs, which in the past felt powerless to halt illegal exports.[3] In the longterm, the convention should put forth a global ban of transboundary waste movements and allow exceptions only in exceptional cases.

3. ESTABLISHMENT OF A GLOBAL INFORMATION SYSTEM

The Protocol should develop a unified information system and methodology (see also Rec. 6) in order to collect and evaluate the information contained in the PIC procedure and the manifest system, which combined must provide sufficient information for environmentally sound disposal. Effective use of this information will require cooperation in eduction and technical training.

There should be a worldwide data bank on hazardous and other wastes to gather information concerning disposal technologies and options, as well as inventory methodologies and data collection on domestic waste generation. The European Data System on Waste Management (EWADAT)[4] and the UNEP-developed global environmental data bank INFOTERRA[5] could be the basis for it and could be expanded.

Finally, provisions must be made that information relating to the whole process of generation to final disposal of hazardous wastes, including options for waste minimization and source reduction, is disclosed to the public. This is an important condition for the publics active participation in and acceptance of hazardous waste policies.

---

[3]The purpose is to eliminate a loophole contained in the Basel Convention that allows bilateral agreements that are more difficult to monitor by the institution that monitors the international Protocol. Furthermore, bilateral agreements represent a range of opportunities to engage in illegal HW traffic through bribes and other illegal activities that have been used to set up HW export schemes in the past. Thus, the Protocol will serve to minimize illegal exports and imports of HWs.

[4]Besides EWADAT, several European countries have already developed environmental data banks, such as FBR, UFORDAT, and so on.

[5]INFOTERRA is part of UNEP's three-part EARTHWATCH system, which further includes the Global Environment Monitoring System (GEMS) and the International Register for Potentially Toxic Chemicals (IRPTC). INFOTERRA is a mechanism to exchange environmental information and could, in combination with IRPTC, provide a comprehensive information system on hazardous and nonhazardous wastes.

4. INITIATION OF FUNDING MECHANISMS FOR HWM TECHNOLOGY TRANSFER

The transfer of hazardous waste management technologies is another crucial input for LLDCs and NICs to manage their own wastes in as environmentally sound a manner as possible. However, the more advanced the technology and thus the lower the residual risks, the higher its cost is, representing a serious constraint on many countries with little foreign exchange. To assist these countries, industrialized countries and transnational corporations must make efforts to assist this technology transfer. Furthermore, new mechanisms must be found to assist the transfer of licenses and patent-protected technologies.[6] International organizations such as the United Nations Organization for Industrial Development (UNIDO) and the United Nations Centre for Transnational Corporations (UNCTC) should play a key role in coordinating and taking the lead in technology transfer.[7]

The Protocol must also include a code of practice for technology transfer in other industries. Guidelines must be developed to assist LLDCs and NICs in the evaluation of the usefulness and of the potential environmental risks of the technologies to be transferred. The purpose is to eliminate the continued transfer of obsolete or hazardous technologies which can no longer meet the environmental standards of industrialized countries.[8]

5. ESTABLISHMENT OF AN INTERNATIONAL EDUCATION AND TECHNICAL TRAINING CENTER FOR HWM

A crucial issue for any control and monitoring mechanism is the verification of information in the manifest with the cargo of the actual shipment in order to ensure that only specified wastes reach a proper HWMF. This verification process requires well-trained personnel and sophisticated equipment. Customs officers who will generally perform these tasks must, therefore, be trained in waste analysis procedures, as well as in occupational safety.

---

[6] A similar proposal has been discussed for the transfer of non-CFC technologies. Many LLDCs and NICs are either forced to buy CFCs or build CFC-plants to meet their demand. The objective to globally reduce the production and use of CFCs, however, requires mechanisms to meet the demand through other means. The availability of non-CFC technologies, which are highly expensive and often unaffordable to LLDCs and NICs, is crucial for a solution, and instruments have to be found to make these technologies available. An international fund that could be used to buy licences and patents and allow the latter countries to use them for the production of CFC substitutes was discussed. This proposal could be a model for the transfer of highly expensive HWM technologies.

[7] These organizations have in-depth experience with the problems involved in technology transfer, and thus could provide a beneficial focal point for both industrial and developing countries.

[8] For a discussion of policy issues in technology transfer to developing countries, see Ashford and Ayers 1985; Castleman 1979.

The training should include institutional training, counterpart training, short-term courses, conferences, and workshops. Training and education also serve the purpose of transmitting and exchanging valuable information needed in other parts of the world.9 The establishment of opportunities for education and training on an international level will contribute to the protocol's implementation on the national levels.

6. ESTABLISHING COMPREHENSIVE LIABILITY AND INSURANCE RULES

The Protocol should provide provisions for strict liability for both property and third party damage reflecting the concept of the Polluter Pays Principle, as well as the notion that pollution prevention is preferable to pollution abatement. Strict liability also emphasizes the position that the costs of hazardous waste management belong to the waste generator and not to the communities that live near disposal sites (World Bank 1989, 264-67). Furthermore, combined with liability, the Protocol should include a provision that the residual risks inherent in hazardous waste disposal are assessed and insured. Insurance requirements, as they exist primarily in exporting countries, should be extended to cover transport and final disposal of the wastes. It is recommended that a comprehensive system be established from the point when a shipment leaves national borders to its final disposal.

Environmentally sound management of hazardous wastes must be based on the premise that waste generators are responsible for the management of the waste from its generation to disposal. The structure of existing laws in many industrialized countries assigns responsibility and liability to waste generators from waste generation to its domestic disposal. If hazardous wastes are transported abroad, an international system assigns liability to cover the transport and potential accidents. However, if hazardous wastes are moved to another country for final disposal, no international system regulates liability, leaving it up to the importing countrys national laws. As a minimum, this part of the waste life cycle in transboundary movements of hazardous wastes must be covered by the Protocol.

7. ESTABLISHING AND FOSTERING MECHANISMS FOR ENFORCEMENT AND COMPLIANCE

Penalties and other mechanisms for compliance and enforcement must be established in the Protocol in order to avoid the problems resulting from EC policy,

---

9The UNEP Industry and Environment Office, as well as other UN agencies, have successfully been sponsoring and organizing such activities. However, their laudable efforts are limited due to scarce resources and are not enough to keep up with the rapid industrial changes in some parts of the world. See also Biswas 1987.

where noncompliance with issued Directives on hazardous wastes had practically no consequences. The Secretariat and the Conference of the Party, as established by the Basel Convention, must be assigned adequate authority and power to effectively implement mechanisms such as incentives and penalties[10] to ensure compliance with the Protocol. These mechanisms will contribute to the effective enforcement of numerous details, such as information exchange, the manifest, annual reports, and so forth, as spelled out in the Basel Convention. Furthermore, procedures for dispute resolution are equally important in order to ensure environmental protection in case parties to the Protocol disagree on their duties or responsibilities.[11]

8. ESTABLISHMENT OF FINANCIAL ASSISTANCE MECHA-NISMS TO AID LLDCs AND NICs IN IMPLEMENTING THE PROTOCOL

The economic problems in LLDCs and NICs call for provisions of financial assistance, such as "International Finance Transfers" (IFT).[12] Financial assistance is decisive for LLDCs and NICs to establish national hazardous waste management systems and to successfully implement and enforce the Protocol. The result will be increased environmental protection, which concurrently will positively affect the overall process of development, thus serving a double purpose.

A major issue that has been neglected in all previous negotiations, but that is directly related to hazardous waste exports, is the tremendous debt-burden of many LLDCs and NICs. The export of hazardous wastes is only a symptom of the more fundamental problem of economic inequalities, which manifest themselves in the

---

[10]Incentives such as tax rebates, low interest rates, and/or subsidies, as well as penalties such as fines imposed for noncompliance, should be set to achieve environmental protection while effectively supporting a change in economic/environmental attitudes. At the same time, these mechanisms provide finance for improved environmental protection. See also Mahajan 1987.

[11]Since HWs have a negative economic value, there is an incentive to abandon the wastes during the transport (e.g. to dump them into the sea (which has occurred) and thus save disposal costs). In order to back up control of waste shipments, a system of financial security was discussed. This financial security would be paid out automatically to the competent authority by the waste carrier in the event of the loss of the waste. As a result, this system would strengthen the enforcement and could be seen as a complement to penalties. For a discussion, see Smits 1985.

[12]International Finance Transfers (IFT) is a mechanism through which payments are made by one country to another in return for improving prevention and control of transboundary pollution. To ensure that the financial assistance, as called for by the Protocol, is used to develop a HWM system and implement the Protocol, the IFT model could be used. For an analysis of IFT, see OECD 1981.

foreign debts of these countries and the existing global economic order. An effective international protocol may halt the uncontrolled stream of hazardous wastes from North to South, but will not contribute to solving the root of the problem. Therefore, an open dialogue must begin to address the complex issues of the terms of trade and the existing world economic order.

9. ESTABLISHING NATIONAL HW POLICIES, INCLUDING SOURCE REDUCTION STRATEGIES

A requirement of the International Protocol and a necessity for its successful implementation is the establishment of national hazardous waste policies as a high national priority. Countries without a national waste policy are condemned to suffer continued environmental degradation without effective control mechanisms for domestic hazardous waste disposal or for legally or illegally imported wastes. A political program for hazardous waste management must require more than just safe waste management, but must take a holistic approach resulting in consideration of all alternatives to waste generation, such as waste minimization and source reduction.

Waste minimization is the key strategy to eliminate waste-related problems. It involves the overall reduction, recycling, separation, and treatment of hazardous wastes to reduce their volume and toxicity. However, this issue is not only a matter of technological priorities. Given the increasingly visible environmental consequences of human activities involving waste generation, the question of whether the benefits of these activities still outweigh the risks and potential environmental harm becomes legitimate. Development in recent years and especially the practice of uncontrolled and sometimes illegal exports of hazardous wastes to LLDCs and NICs suggests that we are reaching a limit. The absence of acceptable policies to deal with increasing volumes of hazardous wastes necessitates an aggressive policy of source reduction.

Existing legislation and regulations in this regard are insufficient. In the Single European Act of 1986, the EC established four principles for its environmental policy: (1) prevention of pollution is preferred over reparation, (2) pollution must be controlled at the source, (3) the polluter must pay for the costs of control, and (4) environmental considerations must be integrated into other Community policies (Koppen 1988). The first three of these principles express only what should be self-evident. However, they are inefficiently enforced when they are enforced at all. Instead, what is needed are strict laws and strategies that require the application of technologies resulting in as little waste generation as possible. In order to implement successfully such programs, both economic incentives and penalties are necessary. To support such activities, hazardous waste exports should unilaterally be prohibited or severely restricted.[13] As a basic principle, a permit to export should be granted only after all techniques for reuse, recycling, and other forms of domestic disposal are optimally utilized. To carry out this proposition, the international protocol should include a black list of substances and wastes that must not be exported.[14]

10. DEVELOPMENT OF SPECIFIC LEGISLATION AND STANDARDS

The establishment of national hazardous waste policies depends on a legal system, as well as on other socioeconomic factors. Countries without hazardous waste laws have to be assisted in developing legislation and regulations covering definitions and standards for generation, transport, and disposal of hazardous wastes, including standards for public and occupational safety and health. In order to accomplish this, the extensive experience of industrialized countries must be integrated in the international Protocol and result in provisions to cooperate with countries that are in the process of promulgating waste laws. This will also contribute to a global harmonization of regulations, an important step for effective global control. Much progress has been made in this regard (e.g., the development of the IWIC, which provides a unique method of waste identification). To implement the single tasks of this recommendation, active institutional support to LLDCs and NICs is necessary and must be agreed upon in the Protocol.

Industrialized countries' existing hazardous waste control systems also have to be improved. As shown earlier, national classification systems proved to be insufficient, existing regulations were ambiguous, and lack of coordination contributed to ineffective control of hazardous wastes. Thus, an international protocol must set procedures to harmonize national legislation and spell out clear

---

[13]HW export restrictions must include the prohibition or tighter limits on sea dumping and ocean incineration; otherwise, these options could represent loopholes to circumvent the restrictions. The export of HWs should be allowed only if the country of import possesses the legal, administrative, and technical capacity to manage and dispose of the wastes in an environmentally sound manner, to be defined by the Protocol. See *Declaration of 35 States and the EC* in the Final Act of the Basel Convention.

[14]Such a "black list" of substances and wastes was, for example, established in the London Dumping Convention. Substances and wastes on this list must not be dumped in the sea. See also European Environment Review 1988, 36.

[15]In the EC, the harmonization of environmental and, in particular, hazardous waste regulations has been a difficult and slow process. However, the Basel Convention has shown that harmonizing certain procedures is crucial and could be agreed upon, (e.g., PIC, manifest, and the supply of information regarding the waste shipment).

definitions for "environmentally sound waste disposal," and so forth, which could not be solved satisfactorily under previous policies.[15] One of the many tasks of such a legal system is to specify a competent authority and grant it the right and responsibility to stop a hazardous waste shipment if it has reason to believe that the wastes in question will not be disposed of according to the provisions of the Protocol. Legal definitions are also needed to apply effective environmental law for court action and for dispute resolution. In sum, specific hazardous waste legislation will benefit LLDCs and NICs in their development of hazardous waste management systems and improve existing systems in industrialized countries.

11. DEVELOPMENT OF COMPREHENSIVE HAZARDOUS WASTE MANAGEMENT SYSTEMS IN ALL COUNTRIES

The next step for control and environmentally sound management of the residual hazardous wastes requires the establishment of a global hazardous waste management system in all countries that do not yet have one. This requires active cooperation among the parties involved. On an international level, the willingness to cooperate is an essential precondition for successful implementation of national hazardous waste management policies. On a project level, only active cooperation between waste generators and waste disposers will ensure environmentally sound waste disposal "from cradle to grave" (Johnsson 1985, 161). Beginning at the analysis of the waste, through the various control mechanisms, such as the manifest, up to the assessment of the disposal site or facility, close cooperation is essential for environmentally sound disposal and may prevent irrational decisions and potential accidents.

Given the state of the environment in developing and newly industrialized countries, the development of comprehensive hazardous waste management systems is an important step to be able to manage the growing volumes of wastes in these countries. The conditions of LLDCs, such as the preponderance of small-scale industries, must be taken into account. In newly industrialized countries, attention must be given to expanding domestic industries, the role of transnational corporations, and the existence of Export Processing Zones. Furthermore, the development of comprehensive hazardous waste management systems also has repercussions for many industrialized countries where high proportions of hazardous wastes still escape control.[16]

---

[16]For example in the Netherlands, an estimated 10%, in Denmark up to 30% of the wastes are not controlled. Some countries are trying to meet this problem by making connection to, and use of public disposal compulsory, for instance in Bavaria (FRG). Furthermore, West Germany considers the setting up of a country-wide connection and disposal requirement. See (STOA 1989, 14).

These 11 recommendations, if adopted, will result in improvements of the Basel Convention so that effective control and monitoring of transboundary waste movements can be exercised. Furthermore, in the long-term, their implementation will reduce and resolve the waste treatment and disposal problem, as waste minimization and source reduction efforts take effect.

The nature of the recommendations will require different levels of commitment, and their implementation will have different consequences. Some of the recommendations will probably not encounter serious resistance from nation-states. However, there may be more resistance to redefining national sovereignty rights concerning environmental issues and the establishment of liability and insurance rules (Recommendation 6). These two recommendations are, however, the most important as the former will delegate national responsibility and jurisdictional power to a supranational organization, and as the latter will establish clear rules of responsibility in case of damage. Both will have serious consequences for nation-states and therefore initially will be opposed. Nevertheless, their adoption is crucial in establishing an effective policy to control transboundary movements of toxic waste. Ultimately, it may require serious public pressure to effectuate their enactment.

Epilogue

The Basel Convention, as the most advanced international approach to control hazardous waste movements, may have serious consequences for hazardous waste management worldwide. The ambiguous parts of the Convention may eventually be clarified and possibly be agreed upon. However, the policy itself, as it is formulated, may ultimately lead to large-scale exports of HWMFs to the Third World in order to manage hazardous wastes from industrialized countries. The only condition that the Basel Convention poses on this likely scenario is that the application of regulations and standards in LLDCs and NICs be equally as stringent as those in the industrialized countries. This represents a very attractive solution to the waste problems many industrialized countries face *today*, and the very reasons that led to hazardous waste exports in the 1980s may also lead to the export of entire hazardous waste management systems in the 1990s. However, this development should ultimately be halted. It is not an acceptable long-term solution, and, eventually, the hazardous waste management capacity in LLDCs and NICs will be exhausted. Even if LLDCs and NICs are willing to engage in such contracts, there is no scientific consensus that available disposal technologies present "no danger to human health and the environment" (Puckett 1989). This view may also have been anticipated by the World Bank, which has indicated that it will not assist development projects in Third World countries targeted to manage imported hazardous wastes from industrialized countries (Piddington 1989). This, however, is certainly not a guarantee that it will not happen anyway. In fact, there is reason to assume that this scenario will become reality and that even more wastes be shipped from North to South or from West to East. Eventually, a global response focussing on source reduction will need to replace the current preoccupation with waste disposal and treatment.

References

Ashford, N.A, and C. Ayers. 1985. Policy issues for consideration in transferring technology to developing countries. *Ecology Law Quarterly* 12(4):871–905.

Biswas, A.K. 1987. Environmental aspects of hazardous waste management for developing countries: problems and prospects. In *Hazardous Waste Management* ed. S. Maltezou et al., UNIDO, pp. 265–271. London: Tycooly.

Castleman, B.I. 1979. The export of hazardous factories to developing nations. *International Journal of Health Services* 9(4):569–600.

European Environment Review. 1988. European's alliance's 10-point programme. *European Environment Review* 2(2).

International Protection of the Environment. 1975. Declaration of the United Nations Conference on the Human Environment. In *International Protection of the Environment*, ed. B. Rüster and B. Simma, pp. 118–21.

Johnsson, G. 1985. The operation of a waste disposal facility which accepts foreign hazardous waste. In *Transfrontier Movements of Hazardous Wastes*. Paris, pp. 156–184. OECD.

Koppen, I. 1988. *The European Communitys Environment Policy*. European University Institute Working Paper No. 88/328. Florence: EUI.

Mahajan, S.P. 1987. Hazardous waste management in India: policy issues and problems. In *Hazardous Waste Management*. ed. S. Maltezou et al., UNIDO, pp. 287–291. London: Tycooly.

OECD. 1981. Possible role of international financial transfer (IFT) in preventing and controlling transfrontier pollution. In *Transfrontier Pollution and the Role of States*. Report of the (OECD) Secretariat. Paris: OECD.

Piddington, K.W. 1989. Sovereignty and the environment. *Environment* 31(7):18-20, 35–39.

Puckett, J. 1989. *Open Letter To The Delegates Of The Annual Ministerial Conference Of*

The Organization Of African States (OAU). July 17, 1989. Addis Ababa Ethiopia: Greenpeace International.

Roelants du Vivier, F. 1988. *Les Vaisseaux Du Poison: La Route Des Déchets Toxiques.* Paris: Sang de la Terre.

Scientific & Technological Options Assessment Programme. 1989. *Hazardous Waste Management Beyond 1992.* Technical Workshop by the Scientific and Technological Options Assessment. Luxembourg: European Parliament.

Smits, H. 1985. Financial security as a means to improve control of transfrontier movements of hazardous waste. In *Transfrontier Movements of Hazardous Wastes.* pp. 185–196. Paris: OECD. 1985.

World Bank. 1989. *The Safe Disposal of Hazardous Wastes: The Special Needs an Problems of Developing Countries.* ed. R. Batstone et al. Vol. 1. Washington, D.C.: World Bank.

Acronyms

ACP	African, Caribbean and Pacific Countries
CEC	Commssion of the European Community
CEFIC	Conseil Europenne des Federations de L'Industries Chimique
CERCLA	Comprehensive Environment, Response, Conservation and Liablity Act
EC/EEC	European Community, former European Economic Community
ECU	European Currency Unit
EIA	Environmental Impact Assessment
EPA	Environmental Protection Agency (U.S.)
EPA-NEIC	National Enforcement and Investigation Center
EPA-OIA	Office of International Activities
EPZ	Export Processing Zone
GATT	General Agreement on Tariffs and Trade
GEMS	Global Environmental Monitoring System
GNP	Gross National Product
HW	Hazardous Waste
HWM	Hazardous Waste Management
HWMF	Hazardous Waste Management Facility
HSWA	Hazardous Substances Waste Amendments
IAEA	International Atomic Energy Agency
IFT	International Finance Transfer
IMO	International Maritime Organization
IRPTC	International Register of Potentially Toxic Chemicals
IWIC	International Waste Identification Code
LLDC	Low and Least Developed Country
NIC	Newly Industrialized Country
OECD	Organization of Economic Cooperation and Development
PCB	Polychlorinated Byphenyl

PIC	Prior Informed Consent
PPP	Polluter Pays Principle
RCRA	Resource Conservation and Recovery Act
TNC	Transnational Corporation
SEA	Single European Act
UN	United Nations
UNECA	United Nations Economic Commission on Africa
UNEP	United Nations Environment Programme
UNIDO	United Nations Industrial Development Organization
UNCTC	United Nations Centre of Transnational Corporations
UNCTAD	United Nations Conference on Trade and Development

Index